CEOE
Field 11

OSAT
Advanced
Mathematics
Teacher Certification Exam

By: Sharon Wynne, M.S.
Southern Connecticut State University

"And, while there's no reason yet to panic, I think it's only prudent that we make preparations to panic."

XAMonline, INC.
Boston

Library of Congress Cataloging-in-Publication Data

Wynne, Sharon A.
    OSAT Advanced Mathematics Field 11: Teacher Certification / Sharon A. Wynne. -2nd ed.
    ISBN 987-1-58197-646-5
        1. OSAT Advanced Mathematics Field 11.   2. Study Guides.   3. CEOE
        4. Teachers' Certification & Licensure.   5. Careers

Disclaimer:
The opinions expressed in this publication are the sole works of XAMonline and were created independently from the National Education Association, Educational Testing Service, or any State Department of Education, National Evaluation Systems or other testing affiliates.

Between the time of publication and printing, state specific standards as well as testing formats and website information may change that is not included in part or in whole within this product. Sample test questions are developed by XAMonline and reflect similar content as on real tests; however, they are not former tests. XAMonline assembles content that aligns with state standards but makes no claims nor guarantees teacher candidates a passing score. Numerical scores are determined by testing companies such as NES or ETS and then are compared with individual state standards. A passing score varies from state to state.

Printed in the United States of America   œ-1

CEOE: OSAT Advanced Mathematics Field 11
ISBN: 978-1-58197-646-5

About the Subject Assessments

CEOE™: Subject Assessment in the Advanced Mathematics examination

Purpose: The assessments are designed to test the knowledge and competencies of prospective secondary level teachers. The question bank from which the assessment is drawn is undergoing constant revision. As a result, your test may include questions that will not count towards your score.

Test Version: There are two versions of subject assessment for Mathematics in Oklahoma. The Middle Level-Intermediate Mathematics (115) exam emphasizes comprehension in Mathematical Processes and Number Sense; Relations, Functions, and Algebra; Measurement and Geometry; Probability, Statistics, and Discrete Mathematics. The Advanced Mathematics (011) exam emphasizes comprehension in Mathematical Processes and Number Sense; Relations, Functions, and Algebra; Measurement and Geometry; Probability, Statistics, and Discrete Mathematics. The Advanced Mathematics study guide is based on a typical knowledge level of persons who have completed a *bachelor's degree program* in Mathematics.

Time Allowance, Format and Scoring: You will have 4 hours to finish the exam. There are approximately 80 multiple-choice questions and one constructed-response question in the exam. 85% of your total score consists of selected-response questions; 15% of your total score consists of the constructed response question

Weighting: There is one constructed-response question in Relation, Functions, and Algebra.

Additional Information about the CEOE Assessments: The CEOE series subject assessments are developed by *National Evaluation Systems.* They provide additional information on the CEOE series assessments, including registration, preparation and testing procedures and study materials such topical guides that have about 30 pages of information including approximately 11 additional sample questions.

Table of Contents

Great Study and Testing Tips!

What to study in order to prepare for the subject assessments is the focus of this study guide but equally important is *how* you study.

You can increase your chances of truly mastering the information by taking some simple, but effective steps.

Study Tips:

1. Some foods aid the learning process. Foods such as milk, nuts, seeds, rice, and oats help your study efforts by releasing natural memory enhancers called CCKs (*cholecystokinin*) composed of *tryptopha*n, *choline*, and *phenylalanine*. All of these chemicals enhance the neurotransmitters associated with memory. Before studying, try a light, protein-rich meal of eggs, turkey, and fish. All of these foods release the memory enhancing chemicals. The better the connections, the more you comprehend.

Likewise, before you take a test, stick to a light snack of energy boosting and relaxing foods. A glass of milk, a piece of fruit, or some peanuts all release various memory-boosting chemicals and help you to relax and focus on the subject at hand.

2. Learn to take great notes. A by-product of our modern culture is that we have grown accustomed to getting our information in short doses (i.e. TV news sound bites or USA Today style newspaper articles.)

Consequently, we've subconsciously trained ourselves to assimilate information better in neat little packages. If your notes are scrawled all over the paper, it fragments the flow of the information. Strive for clarity. Newspapers use a standard format to achieve clarity. Your notes can be much clearer through use of proper formatting. A very effective format is called the *"Cornell Method."*

Take a sheet of loose-leaf lined notebook paper and draw a line all the way down the paper about 1-2" from the left-hand edge.

Draw another line across the width of the paper about 1-2" up from the bottom. Repeat this process on the reverse side of the page.

Look at the highly effective result. You have ample room for notes, a left hand margin for special emphasis items or inserting supplementary data from the textbook, a large area at the bottom for a brief summary, and a little rectangular space for just about anything you want.

3. <u>Get the concept then the details</u>. Too often we focus on the details and don't gather an understanding of the concept. However, if you simply memorize only dates, places, or names, you may well miss the whole point of the subject.

A key way to understand things is to put them in your own words. If you are working from a textbook, automatically summarize each paragraph in your mind. If you are outlining text, don't simply copy the author's words.

Rephrase them in your own words. You remember your own thoughts and words much better than someone else's, and subconsciously tend to associate the important details to the core concepts.

4. <u>Ask Why?</u> Pull apart written material paragraph by paragraph and don't forget the captions under the illustrations.

Example: If the heading is "Stream Erosion", flip it around to read "Why do streams erode?" Then answer the questions.

If you train your mind to think in a series of questions and answers, not only will you learn more, but it also helps to lessen the test anxiety because you are used to answering questions.

5. <u>Read for reinforcement and future needs</u>. Even if you only have 10 minutes, put your notes or a book in your hand. Your mind is similar to a computer; you have to input data in order to have it processed. *By reading, you are creating the neural connections for future retrieval.* The more times you read something, the more you reinforce the learning of ideas.

Even if you don't fully understand something on the first pass, *your mind stores much of the material for later recall.*

6. <u>Relax to learn so go into exile</u>. Our bodies respond to an inner clock called biorhythms. Burning the midnight oil works well for some people, but not everyone.

If possible, set aside a particular place to study that is free of distractions. Shut off the television, cell phone, pager and exile your friends and family during your study period.

If you really are bothered by silence, try background music. Light classical music at a low volume has been shown to aid in concentration over other types. Music that evokes pleasant emotions without lyrics are highly suggested. Try just about anything by Mozart. It relaxes you.

7. Use arrows not highlighters. At best, it's difficult to read a page full of yellow, pink, blue, and green streaks. Try staring at a neon sign for a while and you'll soon see that the horde of colors obscure the message.

A quick note, a brief dash of color, an underline, and an arrow pointing to a particular passage is much clearer than a horde of highlighted words.

8. Budget your study time. Although you shouldn't ignore any of the material, *allocate your available study time in the same ratio that topics may appear on the test.*

Testing Tips:

1. <u>Get smart, play dumb</u>. Don't read anything into the question. Don't make an assumption that the test writer is looking for something else than what is asked. Stick to the question as written and don't read extra things into it.

2. <u>Read the question and all the choices *twice* before answering the question</u>. You may miss something by not carefully reading, and then re-reading both the question and the answers.

If you really don't have a clue as to the right answer, leave it blank on the first time through. Go on to the other questions, as they may provide a clue as to how to answer the skipped questions.

If later on, you still can't answer the skipped ones . . . *Guess.* The only penalty for guessing is that you *might* get it wrong. Only one thing is certain; if you don't put anything down, you will get it wrong!

3. <u>Turn the question into a statement</u>. Look at the way the questions are worded. The syntax of the question usually provides a clue. Does it seem more familiar as a statement rather than as a question? Does it sound strange?

By turning a question into a statement, you may be able to spot if an answer sounds right, and it may also trigger memories of material you have read.

4. <u>Look for hidden clues</u>. It's actually very difficult to compose multiple-foil (choice) questions without giving away part of the answer in the options presented.

In most multiple-choice questions you can often readily eliminate one or two of the potential answers. This leaves you with only two real possibilities and automatically your odds go to Fifty-Fifty for very little work.

5. <u>Trust your instincts</u>. For every fact that you have read, you subconsciously retain something of that knowledge. On questions that you aren't really certain about, go with your basic instincts. **Your first impression on how to answer a question is usually correct.**

6. <u>Mark your answers directly on the test booklet</u>. Don't bother trying to fill in the optical scan sheet on the first pass through the test.

Just be very careful not to miss-mark your answers when you eventually transcribe them to the scan sheet.

7. <u>Watch the clock</u>! You have a set amount of time to answer the questions. Don't get bogged down trying to answer a single question at the expense of 10 questions you can more readily answer.

THIS PAGE BLANK

SUBAREA I.   MATHEMATICAL PROCESSES AND NUMBER SENSE

COMPETENCY 0001   UNDERSTAND MATHEMATICAL PROBLEM
                       SOLVING AND THE CONNECTIONS BETWEEN
                       AND AMONG THE FIELDS OF MATHEMATICS
                       AND OTHER DISCIPLINES.

Successful math teachers introduce their students to multiple problem solving strategies and create a classroom environment where free thought and experimentation are encouraged. Teachers can promote problem solving by allowing multiple attempts at problems, giving credit for reworking test or homework problems, and encouraging the sharing of ideas through class discussion. There are several specific problem solving skills with which teachers should be familiar.

The **guess-and-check** strategy calls for students to make an initial guess at the solution, check the answer, and use the outcome of to guide the next guess. With each successive guess, the student should get closer to the correct answer. Constructing a table from the guesses can help organize the data.

Example:

There are 100 coins in a jar. 10 are dimes. The rest are pennies and nickels. There are twice as many pennies as nickels. How many pennies and nickels are in the jar?

There are 90 total nickels and pennies in the jar (100 coins – 10 dimes).

There are twice as many pennies as nickels. Make guesses that fulfill the criteria and adjust based on the answer found. Continue until we find the correct answer, 60 pennies and 30 nickels.

Number of Pennies	Number of Nickels	Total Number of Pennies and Nickels
40	20	60
80	40	120
70	35	105
60	30	90

When solving a problem where the final result and the steps to reach the result are given, students must **work backwards** to determine what the starting point must have been.

Example:

John subtracted seven from his age, and divided the result by 3. The final result was 4. What is John's age?

> Work backward by reversing the operations.
> $4 \times 3 = 12$;
> $12 + 7 = 19$
> John is 19 years old.

Estimation and testing for **reasonableness** are related skills students should employ both before and after solving a problem. These skills are particularly important when students use calculators to find answers.

Example:

Find the sum of $4387 + 7226 + 5893$.

$4300 + 7200 + 5800 = 17300$	Estimation.
$4387 + 7226 + 5893 = 17506$	Actual sum.

By comparing the estimate to the actual sum, students can determine that their answer is reasonable.

The **questioning technique** is a mathematic process skill in which students devise questions to clarify the problem, eliminate possible solutions, and simplify the problem solving process. By developing and attempting to answer simple questions, students can tackle difficult and complex problems.

The use of supplementary materials in the classroom can greatly enhance the learning experience by stimulating student interest and satisfying different learning styles. Manipulatives, models, and technology are examples of tools available to teachers.

Manipulatives are materials that students can physically handle and move. Manipulatives allow students to understand mathematic concepts by allowing them to see concrete examples of abstract processes. Manipulatives are attractive to students because they appeal to the students' visual and tactile senses. Available for all levels of math, manipulatives are useful tools for reinforcing operations and concepts. They are not, however, a substitute for the development of sound computational skills.

Models are another means of representing mathematical concepts by relating the concepts to real-world situations.

Teachers must choose wisely when devising and selecting models because, to be effective, models must be applied properly. For example, a building with floors above and below ground is a good model for introducing the concept of negative numbers. It would be difficult, however, to use the building model in teaching subtraction of negative numbers.

Finally, there are many forms of **technology** available to math teachers. For example, students can test their understanding of math concepts by working on skill specific computer programs and websites. Graphing calculators can help students visualize the graphs of functions. Teachers can also enhance their lectures and classroom presentations by creating multimedia presentations.

Example – Life Science

Examine an animal population and vegetation density in a biome over time.

Example – Physical Science

Explore motions and forces by calculating speeds based on distance and time traveled and creating a graph to represent the data.

Example – Geography

Explore and illustrate knowledge of earth landforms.

Example – Economics/Finance

Compare car buying with car leasing by graphing comparisons and setting up monthly payment schedules based on available interest rates.

Mathematics dates back before recorded history. Prehistoric cave paintings with geometrical figures and slash counting have been dated prior to 20,000 BC in Africa and France. The major early uses of mathematics were for astronomy, architecture, trading and taxation.

The early history of mathematics is found in Mesopotamia (Sumeria and Babylon), Egypt, Greece and Rome. Noted mathematicians from these times include Euclid, Pythagoras, Apollonius, Ptolemy and Archimedes.

Islamic culture from the 6th through 12th centuries drew from areas ranging from Africa and Spain to India. Through India, they also drew on China. This mix of cultures and ideas brought about developments in many areas, including the concept of algebra, our current numbering system, and major developments in algebra with concepts such as zero. India was the source of many of these developments. Notable scholars of this era include Omar Khayyam and Muhammad al-Khwarizmi.

Counting boards have been found in archeological digs in Babylonia and Greece. These include the Chinese abacus whose current form dates from approximately 1200 AD. Prior to the development of the zero, a counting board or abacus was the common method used for all types of calculations.

Abelard and Fibonacci brought Islamic texts to Europe in the 12th century. By the 17th century, major new works appeared from Galileo and Copernicus (astronomy), Newton and Leibniz (calculus), and Napier and Briggs (logarithms). Other significant mathematicians of this era include René Descartes, Carl Gauss, Pierre de Fermat, Leonhard Euler and Blaise Pascal.

The growth of mathematics since 1800 has been enormous, and has affected nearly every area of life. Some names significant in the history of mathematics since 1800 (and the work they are most known for):

Joseph-Louis Lagrange (theory of functions and of mechanics)
Pierre-Simon Laplace (celestial mechanics, probability theory)
Joseph Fourier (number theory)
Lobachevsky and Bolyai (non-Euclidean geometry)
Charles Babbage (calculating machines, origin of the computer)
Lady Ada Lovelace (first known program)
Florence Nightingale (nursing, statistics of populations)
Bertrand Russell (logic)
James Maxwell (differential calculus and analysis)
John von Neumann (economics, quantum mechanics and game theory)
Alan Turing (theoretical foundations of computer science)
Albert Einstein (theory of relativity)
Gustav Roch (topology)

COMPETENCY 0002   **UNDERSTAND THE PRINCIPLES AND PROCESSES OF MATHEMATICAL REASONING.**

Evaluate Arguments

Conditional statements can be diagrammed using a **Venn diagram**. A diagram can be drawn with one circle inside another circle. The inner circle represents the hypothesis. The outer circle represents the conclusion. If the hypothesis is taken to be true, then you are located inside the inner circle. If you are located in the inner circle then you are also inside the outer circle, so that proves the conclusion is true. Sometimes that conclusion can then be used as the hypothesis for another conditional, which can result in a second conclusion.

Suppose that these statements were given to you, and you are asked to try to reach a conclusion. The statements are:

All swimmers are athletes.
All athletes are scholars.

In "if-then" form, these would be:
If you are a swimmer, then you are an athlete.
If you are an athlete, then you are a scholar.

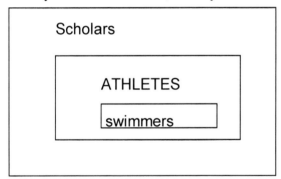

Clearly, if you are a swimmer, then you are also an athlete. This includes you in the group of scholars.

Suppose that these statements were given to you, and you are asked to try to reach a conclusion. The statements are:

All swimmers are athletes.
All wrestlers are athletes.

In "if-then" form, these would be:
    If you are a swimmer, then you are an athlete.
    If you are a wrestler, then you are an athlete.

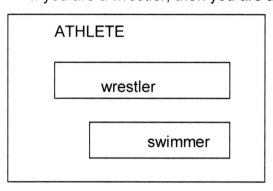

Clearly, if you are a swimmer or a wrestler, then you are also an athlete. This does NOT allow you to come to any other conclusions.

A swimmer may or may NOT also be a wrestler. Therefore, NO CONCLUSION IS POSSIBLE.

Suppose that these statements were given to you, and you are asked to try to reach a conclusion. The statements are:

    All rectangles are parallelograms.
    Quadrilateral ABCD is not a parallelogram.

In "if-then" form, the first statement would be:
    If a figure is a rectangle, then it is also a parallelogram.

Note that the second statement is the negation of the conclusion of statement one. Remember also that the contrapositive is logically equivalent to a given conditional. That is, **"If ⌐ q, then ⌐ p"**. Since "ABCD is NOT a parallelogram " is like saying **"If ⌐ q,"** then you can come to the conclusion **"then ⌐ p"**. Therefore, the conclusion is ABCD is not a rectangle. Looking at the Venn diagram below, if all rectangles are parallelograms, then rectangles are included as part of the parallelograms. Since quadrilateral ABCD is not a parallelogram, it is excluded from anywhere inside the parallelogram box. This allows you to conclude that ABCD cannot be a rectangle either.

PARALLELOGRAMS	
	rectangles

quadrilateral
ABCD

Try These. (The answers are in the Answer Key to Practice Problems):

What conclusion, if any, can be reached? Assume each statement is true, regardless of any personal beliefs.

1. If the Red Sox win the World Series, I will die.
    I died. *The Red Sox won the World series.*

2. If an angle's measure is between 0° and 90°, then the angle is acute. Angle B is not acute.
    Angle B's measure is not between 0° & 90°.

3. Students who do well in geometry will succeed in college.
    Annie is doing extremely well in geometry.
    Annie will succeed in college.

4. Left-handed people are witty and charming.
    You are left-handed.
    I am witty & charming.

Algebraic Postulates for Proofs

The following algebraic postulates are frequently used as reasons for statements in 2 column geometric proofs:

Addition Property:

If $a = b$ and $c = d$, then $a + c = b + d$.

Subtraction Property:

If $a = b$ and $c = d$, then $a - c = b - d$.

Multiplication Property:

If $a = b$ and $c \neq 0$, then $ac = bc$.

Division Property:

If $a = b$ and $c \neq 0$, then $a/c = b/c$.

Reflexive Property:	$a = a$
Symmetric Property:	If $a = b$, then $b = a$.
Transitive Property:	If $a = b$ and $b = c$, then $a = c$.
Distributive Property:	$a(b + c) = ab + ac$
Substitution Property:	If $a = b$, then b may be substituted for a in any other expression (a may also be substituted for b).

Write Direct and Indirect Proofs.

In a 2 column proof, the left side of the proof should be the given information, or statements that could be proved by deductive reasoning. The right column of the proof consists of the reasons used to determine that each statement to the left was verifiably true. The right side can identify given information, or state theorems, postulates, definitions or algebraic properties used to prove that particular line of the proof is true.

Assume the opposite of the conclusion. Keep your hypothesis and given information the same. Proceed to develop the steps of the proof, looking for a statement that contradicts your original assumption or some other known fact. This contradiction indicates that the assumption you made at the beginning of the proof was incorrect; therefore, the original conclusion has to be true.

Estimation and Approximation to Check Reasonableness

Estimation and approximation may be used to check the reasonableness of answers.

<u>Example</u>: Estimate the answer.

$$\frac{58 \times 810}{1989}$$

58 becomes 60, 810 becomes 800 and 1989 becomes 2000.

$$\frac{60 \times 800}{2000} = 24$$

Word problems: An estimate may sometimes be all that is needed to solve a problem.

<u>Example</u>: Janet goes into a store to purchase a CD on sale for $13.95. While shopping, she sees two pairs of shoes, prices $19.95 and $14.50. She only has $50. Can she purchase everything? (Assume there is no sales tax.)[SA1]

Solve by rounding:

$19.95 → $20.00
$14.50 → $15.00
$13.95 → <u>$14.00</u>
   $49.00   Yes, she can purchase the CD and the shoes.

Apply inductive and deductive reasoning to solve problems.

Inductive thinking is the process of finding a pattern from a group of examples. That pattern is the conclusion that this set of examples seemed to indicate. It may be a correct conclusion or it may be an incorrect conclusion because other examples may not follow the predicted pattern.

Deductive thinking is the process of arriving at a conclusion based on other statements that are all known to be true, such as theorems, axiomspostulates, or postulates. Conclusions found by deductive thinking based on true statements will **always** be true.

Examples:

Suppose:
On Monday Mr. Peterson eats breakfast at McDonalds.
On Tuesday Mr. Peterson eats breakfast at McDonalds.
On Wednesday Mr. Peterson eats breakfast at McDonalds.
On Thursday Mr. Peterson eats breakfast at McDonalds again.

Conclusion: On Friday Mr. Peterson will eat breakfast at McDonalds again.

This is a conclusion based on inductive reasoning. Based on several days observations, you conclude that Mr. Peterson will eat at McDonalds. This may or may not be true, but it is a conclusion arrived at by inductive thinking.

A **counterexample** is an exception to a proposed rule or conjecture that disproves the conjecture. For example, the existence of a single non-brown dog disproves the conjecture "all dogs are brown". Thus, any non-brown dog is a counterexample.

In searching for mathematic counterexamples, one should consider extreme cases near the ends of the domain of an experiment and special cases where an additional property is introduced. Examples of extreme cases are numbers near zero and obtuse triangles that are nearly flat. An example of a special case for a problem involving rectangles is a square because a square is a rectangle with the additional property of symmetry.

Example:

Identify a counterexample for the following conjectures.

1. If n is an even number, then $n + 1$ is divisible by 3.

   $n = 4$
   $n + 1 = 4 + 1 = 5$
   5 is not divisible by 3.

2. If n is divisible by 3, then $n^2 - 1$ is divisible by 4.

   $n = 6$
   $n^2 - 1 = 6^2 - 1 = 35$
   35 is not divisible by 4.

Proofs by mathematical induction.

Proof by induction states that a statement is true for all numbers if the following two statements can be proven:

1. The statement is true for $n = 1$.
2. If the statement is true for $n = k$, then it is also true for $n = k+1$.

In other words, we must show that the statement is true for a particular value and then we can assume it is true for another, larger value (k). Then, if we can show that the number after the assumed value ($k+1$) also satisfies the statement, we can assume, by induction, that the statement is true for all numbers.

The four basic components of induction proofs are: (1) the statement to be proved, (2) the beginning step ("let $n = 1$"), (3) the assumption step ("let $n = k$ and assume the statement is true for k, and (4) the induction step ("let $n = k+1$").

Example:

Prove that the sum all numbers from 1 to n is equal to $\dfrac{(n)(n+1)}{2}$.

Let $n = 1$.   Beginning step.

Then the sum of 1 to 1 is 1.

And $\dfrac{(n)(n+1)}{2} = 1$.

Thus, the statement is true for $n = 1$.   Statement is true in a particular instance.

Assumption:

Let $n = k + 1$

$k = n - 1$

Then $[1 + 2 + \ldots + k] + (k+1) = \dfrac{(k)(k+1)}{2} + (k+1)$

Substitute the assumption.

$= \dfrac{(k)(k+1)}{2} + \dfrac{2(k+1)}{2}$

Common denominator.

$= \dfrac{(k)(k+1) + 2(k+1)}{2}$

Add fractions.

$= \dfrac{(k+2)(k+1)}{2}$

Simplify.

$= \dfrac{(k+1)+1)(k+1)}{2}$

Write in terms of $k+1$.

For $n = 4$, $k = 3$

$= \dfrac{(4+1)(4)}{2} = \dfrac{20}{2} = 10$

Conclude that the original statement is true for $n = k+1$ if it is assumed that the statement is true for $n = k$.

Proofs on a coordinate plane

Use proofs on the coordinate plane to prove properties of geometric figures. Coordinate proofs often utilize formulas such as the Distance Formula, Midpoint Formula, and the Slope Formula.

The most important step in coordinate proofs is the placement of the figure on the plane. Place the figure in such a way to make the mathematical calculations as simple as possible.

Example:

1. Prove that the square of the length of the hypotenuse of triangle ABC is equal to the sum of the squares of the lengths of the legs using coordinate geometry.

Draw and label the graph.

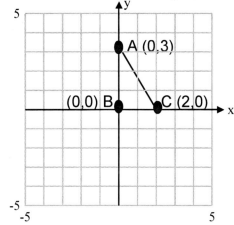

Use the distance formula to find the lengths of the sides of the triangle.

$$d = \sqrt{(x_2 - x_1)^2 + (y_2 - y_1)^2}$$

AB = $\sqrt{3^2} = 3$, BC = $\sqrt{2^2} = 2$, AC = $\sqrt{3^2 + 2^2} = \sqrt{13}$

Conclude

$(AB)^2 + (BC)^2 = 3^2 + 2^2 = 13$
$(AC)^2 = (\sqrt{13})^2 = 13$

Thus, $(AB)^2 + (BC)^2 = (AC)^2$

COMPETENCY 0003   **UNDERSTAND AND COMMUNICATE MATHEMATICAL CONCEPTS AND SYMBOLS.**

Students of mathematics must be able to recognize and interpret the different representations of arithmetic operations.

First, there are many different verbal descriptions for the operations of addition, subtraction, multiplication, and division. The table below identifies several words and/or phrases that are often used to denote the different arithmetic operations.

Operation	Descriptive Words
Addition	"plus", "combine", "sum", "total", "put together"
Subtraction	"minus", "less", "take away", "difference"
Multiplication	"product", "times", "groups of"
Division	"quotient", "into", "split into equal groups",

Students of mathematics must be able to recognize and interpret the different representations of arithmetic operations. Diagrams of arithmetic operations can present mathematical data in visual form.

For example, we can use the number line to add and subtract.

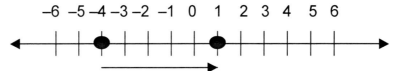

The addition of 5 to – 4 on the number line; – 4 + 5 = 1.

We can also use pictorial representations to explain all of the arithmetic processes.

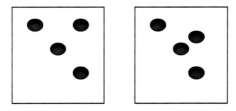

Two groups of four equals eight or 2 x 4 = 8 shown in picture form.

Adding three objects to two or 3 + 2 = 5 shown in picture form.

Examples, illustrations, and symbolic representations are useful tools in explaining and understanding mathematical concepts. The ability to create examples and alternative methods of expression allows students to solve real world problems and better communicate their thoughts.

Examples, illustrations, and symbolic representations are useful tools in explaining and understanding mathematical concepts. The ability to create examples and alternative methods of expression allows students to solve real world problems and better communicate their thoughts.

Concrete examples are real world applications of mathematical concepts.

For example, measuring the shadow produced by a tree or building is a real world application of trigonometric functions, acceleration or velocity of a car is an application of derivatives, and finding the volume or area of a swimming pool is a real world application of geometric principles.

Pictorial illustrations of mathematic concepts help clarify difficult ideas and simplify problem solving.

Examples:

1. Rectangle R represents the 300 students in School A. Circle P represents the 150 students that participated in band. Circle Q represents the 170 students that participated in a sport. 70 students participated in both band and a sport.

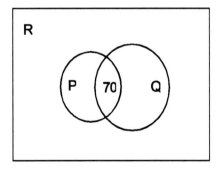

Pictorial representation of above situation.

2. A ball rolls up an incline and rolls back to its original position. Create a graph of the velocity of the ball.

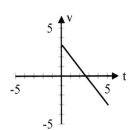

Velocity starts out at its maximum as the ball begins to roll, decreases to zero at the top of the incline, and returns to the maximum in the opposite direction at the bottom of the incline.

Symbolic representation is the basic language of mathematics. Converting data to symbols allows for easy manipulation and problem solving. Students should have the ability to recognize what the symbolic notation represents and convert information into symbolic form. For example, from the graph of a line, students should have the ability to determine the slope and intercepts and derive the line's equation from the observed data.
Another possible application of symbolic representation is the formulation of algebraic expressions and relations from data presented in word problem form.

In order to understand mathematics and solve problems, one must know the definitions of basic mathematic terms and concepts. For a list of definitions and explanations of basic math terms, visit the following website: http://home.blarg.net/~math/deflist.html

Additionally, one must use the language of mathematics correctly and precisely to accurately communicate concepts and ideas.

For example, the statement "minus ten times minus five equals plus fifty" is incorrect because minus and plus are arithmetic operations not numerical modifiers. The statement should read "negative ten times negative five equals positive 50".

Successful teachers are able to identify all types of student errors. Identifying patterns of errors allows teachers to tailor their curriculum to enhance student learning and understanding. The following is a list of common errors committed by algebra[SA2] and calculus students.

Algebra

1. Failing to distribute:
    ex. $4x - (2x - 3) = 4x - 2x + 3$ (**not** $4x - 2x - 3$)
    ex. $3(2x + 5) = 6x + 15$ (**not** $6x + 5$)

2. Distributing exponents:
    ex. $(x + y)^2 = x^2 + 2xy + y^2$ (**not** $x^2 + y^2$)

3. Canceling terms instead of factors:
    ex. $\dfrac{x^3 + 5}{x} \neq x^2 + 5 \quad = \quad x^2 + \dfrac{5}{x}$ (**not** $x^2 + 5$)

4. Misunderstanding negative and fractional exponents:
    ex. $x^{\frac{1}{2}} = \sqrt{x}$ (**not** $\dfrac{1}{x^2}$)
    ex. $x^{-2} = \dfrac{1}{x^2}$ (**not** $\sqrt{x}$)

Calculus

1. Failing to use the product, quotient, and chain rules:
    $f(x) = (5x^2)(3x^3),$
    ex. $f'(x) = (10x)(3x^3) + (5x^2)(9x^2)$
    $= 30x^4 + 45x^4 = 75x^4$
    [**not** $(10x)(9x^2) = 90x^3$]

2. Misapplying derivative and integration rules:
    ex. $\displaystyle\int \frac{1}{x} = \ln x + C,$
    $\displaystyle\int \frac{1}{(x^2 + 1)} \neq \ln(x^2 + 1) + C$

Teachers also must not neglect multiple errors in the work of a student that may be disguised by a correct answer. For example, failing twice to properly distribute negative one in the following arithmetic operation produces a deceptively correct answer.

$4x - (2x+3) - (2x - 3) = 4x - 2x + 3 - 2x - 3 = -2x$ (incorrect distribution)

$4x - (2x+3) - (2x - 3) = 4x - 2x - 3 - 2x + 3 = -2x$ (correct distribution)

COMPETENCY 0004   UNDERSTAND NUMBER THEORY AND PRINCIPLES AND PROPERTIES OF THE REAL AND COMPLEX NUMBER SYSTEMS.

Real Number System

a. **Natural numbers**—the counting numbers, $1, 2, 3, \ldots$

b. **Whole numbers**—the counting numbers along with zero, $0, 1, 2 \ldots$

c. **Integers**—the counting numbers, their opposites, and zero, $\ldots, {}^-1, 0, 1, \ldots$

d. **Rationals**—all of the fractions that can be formed from the whole numbers. Zero cannot be the denominator. In decimal form, these numbers will either be terminating or repeating decimals. Simplify square roots to determine if the number can be written as a fraction.

e. **Irrationals**—real numbers that cannot be written as a fraction. The decimal forms of these numbers are neither terminating nor repeating. Examples: $\pi, e, \sqrt{2}$, etc.

f. **Real numbers**—the set of numbers obtained by combining the rationals and irrationals. Complex numbers, i.e. numbers that involve i or $\sqrt{-1}$ [SA3], are not real numbers.

Denseness Property

The Denseness Property of real numbers states that, if all real numbers are ordered from least to greatest on a number line, there is an infinite set of real numbers between any two given numbers on the line.

Example:

Between 7.6 and 7.7, there is the rational number 7.65 in the set of real numbers.

Between 3 and 4 there exists no other natural number.

Decimals and Fractions

To convert a fraction to a decimal, simply divide the numerator (top) by the denominator (bottom). Use long division if necessary.

If a decimal has a fixed number of digits, the decimal is said to be terminating. To write such a decimal as a fraction, first determine what place value the farthest right digit is in, for example: tenths, hundredths, thousandths, ten thousandths, hundred thousands, etc. Then drop the decimal and place the string of digits over the number given by the place value.

If a decimal continues forever by repeating a string of digits, the decimal is said to be repeating. To write a repeating decimal as a fraction, follow these steps.

a.   Let $x =$ the repeating decimal
    (ex. $x = .716716716...$)

b.   Multiply x by the multiple of ten that will move the decimal just to the right of the repeating block of digits.
    (ex. $1000x = 716.716716...$)

c.   Subtract the first equation from the second.
    (ex. $1000x - x = 716.716.716... - .716716...$)

d.   Simplify and solve this equation. The repeating block of digits will subtract out.
    (ex. $999x = 716$ so $x = {}^{716}\!/_{999}$)

e.   The solution will be the fraction for the repeating decimal.

Order of Operations

The Order of Operations are to be followed when evaluating algebraic expressions. Remember the mnemonic PEMDAS (Please Excuse My Dear Aunt Sally) to follow these steps in order[SA4]:

1. Simplify inside grouping characters such as parentheses, brackets, radicals, fraction bars, etc.

2. Multiply out expressions with exponents.

3. Do multiplication or division from left to right.

4. Do addition or subtraction from left to right.

Samples of simplifying expressions with exponents:

$(-2)^3 = -8$ $-2^3 = -8$

$(-2)^4 = 16$ $-2^4 = -16$ Note change of sign.

$(\tfrac{2}{3})^3 = \tfrac{8}{27}$ [SA5]

$5^0 = 1$

$4^{-1} = \tfrac{1}{4}$

Inverse Operations

Subtraction is the inverse of Addition, and vice-versa.
Division is the inverse of Multiplication, and vice-versa.
Taking a square root is the inverse of squaring, and vice-versa.

These inverse operations are used when solving equations.

A **ratio** is a comparison of 2 numbers. If a class had 11 boys and 14 girls, the ratio of boys to girls could be written one of 3 ways:

$$11:14 \quad \text{or} \quad 11 \text{ to } 14 \quad \text{or} \quad \frac{11}{14}$$

The ratio of girls to boys is:

$$14:11, \ 14 \text{ to } 11 \text{ or} \quad \frac{14}{11}$$

Ratios can be reduced when possible. A ratio of 12 cats to 18 dogs would reduce to 2:3, 2 to 3 or $2/3$.

Note: Read ratio questions carefully. Given a group of 6 adults and 5 children, the ratio of children to the entire group would be 5:11.
A **proportion** is an equation in which a fraction is set equal to another. To solve the proportion, multiply each numerator times the other fraction's denominator. Set these two products equal to each other and solve the resulting equation. This is called **cross-multiplying** the proportion.

Example: $\dfrac{4}{15} = \dfrac{x}{60}$ is a proportion.

To solve this, cross multiply.

$$(4)(60) = (15)(x)$$
$$240 = 15x$$
$$16 = x$$

Example: $\dfrac{x+3}{3x+4} = \dfrac{2}{5}$ is a proportion.

To solve, cross multiply.

$$5(x+3) = 2(3x+4)$$
$$5x + 15 = 6x + 8$$
$$7 = x$$

Example: $\dfrac{x+2}{8} = \dfrac{2}{x-4}$ is another proportion.

To solve, cross multiply.

$$(x+2)(x-4) = 8(2)$$
$$x^2 - 2x - 8 = 16$$
$$x^2 - 2x - 24 = 0$$
$$(x-6)(x+4) = 0$$
$$x = 6 \text{ or } x = {}^-4$$

Both answers work.

Fractions, decimals, and percents can be used interchangeably within problems.

→ To change a percent into a decimal, move the decimal point two places to the left and drop off the percent sign.

→ To change a decimal into a percent, move the decimal two places to the right and add on a percent sign.

→ To change a fraction into a decimal, divide the numerator by the denominator.

→ To change a decimal number into an equivalent fraction, write the decimal part of the number as the fraction's numerator. As the fraction's denominator use the place value of the last column of the decimal. Reduce the resulting fraction as far as possible.

Example: J.C. Nickels has Hunch jeans 1/4 off the usual price of $36.00. Shears and Roadkill have the same jeans 30% off their regular price of $40. Find the cheaper price.

1/4 = .25 so .25(36) = $9.00 off $36 - 9 = $27 sale price

30% = .30 so .30(40) = $12 off $40 - 12 = $28 sale price

The price at J.C Nickels is actually lower.

Integral Exponents

Coefficients are the numbers in front of the variables in a term.

Addition/Subtraction with integral exponents.

-**Like terms** have the same variables raised to the same powers. Like terms can be combined by adding or subtracting their coefficients and by keeping the variables and their exponents the same. Unlike terms cannot be combined.

$$5x^4 + 3x^4 - 2x^4 = 6x^4$$
$$3ab^2 + 5ab + ab^2 = 4ab^2 + 5ab$$

Multiplication with integral exponents.
-When multiplying terms together, multiply their coefficients and add the exponents of the same variables. When a term is raised to a power, raise the term's coefficient to the power outside the parentheses. Then multiply the outside exponent times each of the inside exponents.

$$(3a^7b^5)(^-4a^2b^9c^4) = \,^-12a^9b^{14}c^4$$

$$(^-2x^3yz^7)^5 = \,^-32x^{15}y^5z^{35}$$

Division with integral exponents.
-When dividing any number of terms by a single term, divide or reduce their coefficients. Then subtract the exponent of a variable on the bottom from the exponent of the same variable on the top.

$$\frac{24a^8b^7 + 16a^7b^5 - 8a^6}{8a^6} = 3a^2b^7 + 2ab^5 - 1$$

Negative exponents are usually changed into positive exponents by moving those variables from numerator to denominator (or vice versa) to make the exponents become positive.

$$\frac{^-60a^8b^{-7}c^{-6}d^5e^{-12}}{12a^{-2}b^3c^{-9}d^5e^{-4}} = \,^-5a^{10}b^{-10}c^3e^{-8} = \frac{^-5a^{10}c^3}{b^{10}e^8}$$

$$(6x^{-9}y^3)(^-2x^5y^{-1}) = \,^-12x^{-4}y^2 = \frac{^-12y^2}{x^4}$$

Practice:

1. $(4x^4y^2)(^-3xy^{-3}z^5)$

2. $(100a^6 + 60a^4 - 28a^2b^3) \div (4a^2)$

3. $(3x^5y^2)^4 + 8x^9y^6 + 19x^{20}y^8$

Rational Exponents

When an expression has a rational exponent, it can be rewritten using a radical sign. The denominator of the rational exponent becomes the index number on the front of the radical sign. The base of the original expression goes inside the radical sign. The numerator of the rational exponent is an exponent which can be placed either inside the radical sign on the original base or outside the radical as an exponent on the radical expression. The radical can then be simplified as far as possible.

$$4^{3/2} = \sqrt[2]{4^3} \quad \text{or} \quad \left(\sqrt{4}\right)^3 = \sqrt{64} = 8$$

$$16^{3/4} = \sqrt[4]{16^3} \quad \text{or} \quad \left(\sqrt[4]{16}\right)^3 = 2^3 = 8$$

$$25^{-1/2} = \frac{1}{25^{1/2}} = \frac{1}{\sqrt{25}} = \frac{1}{5}$$

Properties of Real Numbers

The real number properties are best explained in terms of a small set of numbers. For each property, a given set will be provided.

Axioms of Addition

Closure—For all real numbers a and b, $a + b$ is a unique real number.

Associative—For all real numbers a, b, and c, $(a + b) + c = a + (b + c)$.

Additive Identity—There exists a unique real number 0 (zero) such that $a + 0 = 0 + a = a$ for every real number a.

Additive Inverses—For each real number a, there exists a real number $-a$ (the opposite of a) such that $a + (-a) = (-a) + a = 0$.

Commutative—For all real numbers a and b, $a + b = b + a$.

Axioms of Multiplication

Closure—For all real numbers a and b, ab is a unique real number.

Associative—For all real numbers a, b, and c, $(ab)c = a(bc)$.

Multiplicative Identity—There exists a unique nonzero real number 1 (one) such that $1 \cdot a = a \cdot 1 = a$.

Multiplicative Inverses—For each nonzero rel number, there exists a real number $1/a$ (the reciprocal of a) such that $a(1/a) = (1/a)a = 1$.

Commutative—For all real numbers a and b, $ab = ba$.

The Distributive Axiom of Multiplication over Addition

For all real numbers a, b, and c, $a(b + c) = ab + ac$.

Prime and Composite Numbers

Prime numbers are whole numbers greater than 1 that have only 2 factors, 1 and the number itself. Examples of prime numbers are 2,3,5,7,11,13,17, or 19. Note that 2 is the only even prime number.

Composite numbers are whole numbers that have more than 2 different factors. For example 9 is composite because besides factors of 1 and 9, 3 is also a factor. 70 is also composite because besides the factors of 1 and 70, the numbers 2,5,7,10,14, and 35 are also all factors.

Remember that the number 1 is neither prime nor composite.

Prime Factorization

Prime numbers are numbers that can only be factored into 1 and the number itself. When factoring into prime factors, all the factors must be numbers that cannot be factored again (without using 1). Initially numbers can be factored into any 2 factors. Check each resulting factor to see if it can be factored again. Continue factoring until all remaining factors are prime. This is the list of prime factors. Regardless of what way the original number was factored, the final list of prime factors will always be the same.

Example: Factor 30 into prime factors.

Factor 30 into any 2 factors.
$5 \cdot 6$ Now factor the 6.
$5 \cdot 2 \cdot 3$ These are all prime factors.

Factor 30 into any 2 factors.
$3 \cdot 10$ Now factor the 10.
$3 \cdot 2 \cdot 5$ These are the same prime factors even though the original factors were different.

Example: Factor 240 into prime factors.

Factor 240 into any 2 factors.
$24 \cdot 10$ Now factor both 24 and 10.
$4 \cdot 6 \cdot 2 \cdot 5$ Now factor both 4 and 6.
$2 \cdot 2 \cdot 2 \cdot 3 \cdot 2 \cdot 5$ These are prime factors.

This can also be written as $2^4 \cdot 3 \cdot 5$.

GCF and LCM

GCF is the abbreviation for the **greatest common factor**. The GCF is the largest number that is a factor of all the numbers given in a problem. The GCF can be no larger than the smallest number given in the problem. If no other number is a common factor, then the GCF will be the number 1. To find the GCF, list all possible factors of the smallest number given (include the number itself). Starting with the largest factor (which is the number itself), determine if it is also a factor of all the other given numbers. If so, that is the GCF. If that factor does not work, try the same method on the next smaller factor. Continue until a common factor is found. That is the GCF. Note: There can be other common factors besides the GCF.

Example: Find the GCF of 12, 20, and 36.

The smallest number in the problem is 12. The factors of 12 are 1, 2, 3, 4, 6, and 12. The largest factor is 12, but it does not divide evenly into 20. Neither does 6, but 4 will divide into both 20 and 36 evenly. Therefore, 4 is the GCF.

Example: Find the GCF of 14 and 15.

Factors of 14 are 1, 2, 7, and 14. The largest factor is 14, but it does not divide evenly into 15. Neither does 7 or 2. Therefore, the only factor common to both 14 and 15 is the number 1, the GCF.

LCM is the abbreviation for **least common multiple**. The least common multiple of a group of numbers is the smallest number into which all of the given numbers will divide. The least common multiple will always be the largest of the given numbers or a multiple of the largest number.

Example: Find the LCM of 20, 30, and 40.

The largest number given is 40, but 30 will not divide evenly into 40. The next multiple of 40 is 80 (2 x 40), but 30 will not divide evenly into 80 either. The next multiple of 40 is 120. 120 is divisible by both 20 and 30, so 120 is the LCM (least common multiple).

Example: Find the LCM of 96, 16, and 24.

The largest number is 96. The number 96 is divisible by both 16 and 24, so 96 is the LCM.

Divisibility Tests

a. A number is divisible by 2 if that number is an even number (which means it ends in 0,2,4,6 or 8).

1,354 ends in 4, so it is divisible by 2. 240,685 ends in a 5, so it is not divisible by 2.

b. A number is divisible by 3 if the sum of its digits is evenly divisible by 3.

The sum of the digits of 964 is 9+6+4 = 19. Since 19 is not divisible by 3, neither is 964. The digits of 86,514 is 8+6+5+1+4 = 24. Since 24 is divisible by 3, 86,514 is also divisible by 3.

c. A number is divisible by 4 if the number in its last 2 digits is evenly divisible by 4.

The number 113,336 ends with the number 36 in the last 2 columns. Since 36 is divisible by 4, then 113,336 is also divisible by 4.

The number 135,627 ends with the number 27 in the last 2 columns. Since 27 is not evenly divisible by 4, then 135,627 is also not divisible by 4.

d. A number is divisible by 5 if the number ends in either a 5 or a 0.

225 ends with a 5 so it is divisible by 5. The number 470 is also divisible by 5 because its last digit is a 0. 2,358 is not divisible by 5 because its last digit is an 8, not a 5 or a 0.

e. A number is divisible by 6 if the number is even and the sum of its digits is evenly divisible by 3.

4,950 is an even number and its digits add to 18. (4+9+5+0 = 18) Since the number is even and the sum of its digits is 18 (which is divisible by 3), then 4950 is divisible by 6. 326 is an even number, but its digits add up to 11. Since 11 is not divisible by 3, then 326 is not divisible by 6. 698,135 is not an even number, so it cannot possibly be divided evenly by 6.

f. A number is divisible by 8 if the number in its last 3 digits is evenly divisible by 8.

The number 113,336 ends with the 3-digit number 336 in the last 3 places. Since 336 is divisible by 8, then 113,336 is also divisible by 8. The number 465,627 ends with the number 627 in the last 3 places. Since 627 is not evenly divisible by 8, then 465,627 is also not divisible by 8.

g. A number is divisible by 9 if the sum of its digits is evenly divisible by 9.

The sum of the digits of 874 is 8+7+4 = 19. Since 19 is not divisible by 9, neither is 874. The digits of 116,514 is 1+1+6+5+1+4 = 18. Since 18 is divisible by 9, 116,514 is also divisible by 9.

h. A number is divisible by 10 if the number ends in the digit 0. 305 ends with a 5 so it is not divisible by 10. The number 2,030,270 is divisible by 10 because its last digit is a 0. 42,978 is not divisible by 10 because its last digit is an 8, not a 0.

i. Why these rules work.

All even numbers are divisible by 2 by definition. A 2-digit number (with T as the tens digit and U as the ones digit) has as its sum of the digits, T + U. Suppose this sum of T + U is divisible by 3. Then it equals 3 times some constant, K. So, T + U = 3K. Solving this for U, U = 3K - T. The original 2 digit number would be represented by 10T + U. Substituting 3K - T in place of U, this 2-digit number becomes 10T + U = 10T + (3K - T) = 9T + 3K. This 2-digit number is clearly divisible by 3, since each term is divisible by 3. Therefore, if the sum of the digits of a number is divisible by 3, then the number itself is also divisible by 3. Since 4 divides evenly into 100, 200, or 300, 4 will divide evenly into any amount of hundreds. The only part of a number that determines if 4 will divide into it evenly is the number in the last 2 places. Numbers divisible by 5 end in 5 or 0. This is clear if you look at the answers to the multiplication table for 5. Answers to the multiplication table for 6 are all even numbers. Since 6 factors into 2 times 3, the divisibility rules for 2 and 3 must both work. Any number of thousands is divisible by 8. Only the last 3 places of the number determine whether or not it is divisible by 8. A 2 digit number (with T as the tens digit and U as the ones digit) has as its sum of the digits, T + U. Suppose this sum of T + U is divisible by 9. Then it equals 9 times some constant, K. So, T + U = 9K. Solving this for U, U = 9K - T. The original 2-digit number would be represented by 10T + U. Substituting 9K - T in place of U, this 2-digit number becomes 10T + U = 10T + (9K - T) = 9T + 9K. This 2-digit number is clearly divisible by 9, since each term is divisible by 9. Therefore, if the sum of the digits of a number is divisible by 9, then the number itself is also divisible by 9. Numbers divisible by 10 must be multiples of 10 which all end in a zero.

Complex numbers are numbers of the form $a + bi$ where a and b are real numbers and i is the imaginary unit $\sqrt{-1}$. Complex numbers attach meaning to and allow calculations with the square root of negative numbers. Electronics and engineering are two disciplines that use applications of complex numbers. When graphing complex numbers, we plot the real number a on the x-axis (labeled R for real) and b on the y-axis (labeled I for imaginary). The modulus is the length of the vector from the origin to the position of the complex number on the graph. The argument is the angle the vector makes with the horizontal axis R.

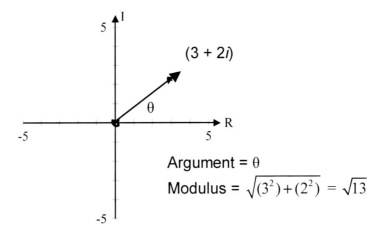

Argument = θ

Modulus = $\sqrt{(3^2)+(2^2)} = \sqrt{13}$

An alternative representation of a complex number is the polar form. We can represent any complex number z with the following equation:

$z = r(\cos \theta + i\sin \theta)$ where r is the modulus and θ is the argument

Writing complex numbers in polar form simplifies multiplication and division applications. Conversely, Cartesian notation ($a + bi$) makes addition and subtraction applications easier to manage.

Complex numbers are of the form $a + b\,\mathbf{i}$, where a and b are real numbers and $i = \sqrt{-1}$. When **i** appears in an answer, it is acceptable unless it is in a denominator. When **i²** appears in a problem, it is always replaced by -1. Remember, **i² = –1**.

To add or subtract complex numbers, add or subtract the real parts. Then add or subtract the imaginary parts and keep the **i** (just like combining like terms).

<u>Examples</u>: Add $(2 + 3i) + (^-7 - 4i)$.

$2 + {}^-7 = {}^-5$ $3i + {}^-4i = {}^-i$ so,

$(2 + 3i) + (^-7 - 4i) = {}^-5 - i$

Subtract $(8 - 5i) - (^-3 + 7i)$

$8 - 5i + 3 - 7i = 11 - 12i$

To multiply 2 complex numbers, F.O.I.L. the 2 numbers together. Replace **i²** with **–1** and finish combining like terms. Answers should have the form $a + b\,\mathbf{i}$.

Example: Multiply $(8+3i)(6-2i)$ F.O.I.L. this.

$48-16i+18i-6i^2$ Let $i^2 = {}^-1$.

$48-16i+18i-6({}^-1)$

$48-16i+18i+6$

$54+2i$ This is the answer.

Example: Multiply $(5+8i)^2$ ← Write this out twice.

$(5+8i)(5+8i)$ F.O.I.L. this

$25+40i+40i+64i^2$ Let $i^2 = {}^-1$.

$25+40i+40i+64({}^-1)$

$25+40i+40i-64$

${}^-39+80i$ This is the answer.

When dividing 2 complex numbers, you must eliminate the complex number in the denominator. If the complex number in the denominator is of the form b **i**, multiply both the numerator and denominator by **i**. Remember to replace i^2 with –1 and then continue simplifying the fraction.

Example:

$$\frac{2+3i}{5i}$$ Multiply this by $\frac{i}{i}$

$$\frac{2+3i}{5i}\times\frac{i}{i}=\frac{(2+3i)\ i}{5i\cdot i}=\frac{2i+3i^2}{5i^2}=\frac{2i+3({}^-1)}{{}^-5}=\frac{{}^-3+2i}{{}^-5}=\frac{3-2i}{5}$$

If the complex number in the denominator is of the form $a+b$ **i**, multiply both the numerator and denominator by **the conjugate of the denominator**. **The conjugate of the denominator** is the same 2 terms with the opposite sign between the 2 terms (the real term does not change signs). The conjugate of $2-3i$ is $2+3i$. The conjugate of $-6+11i$ is $-6-11i$. Multiply together the factors on the top and bottom of the fraction. Remember to replace i^2 with –1, combine like terms, and then continue simplifying the fraction.

Example:

$$\frac{4+7i}{6-5i}$$ Multiply by $\frac{6+5i}{6+5i}$, the conjugate.

$$\frac{(4+7i)}{(6-5i)}\times\frac{(6+5i)}{(6+5i)}=\frac{24+20i+42i+35i^2}{36+30i-30-25i^2}=\frac{24+62i+35(-1)}{36-25(-1)}=\frac{-11+62i}{61}$$

Example:

$$\frac{24}{{}^-3-5i}$$ Multiply by $\dfrac{{}^-3+5i}{{}^-3+5i}$, the conjugate.

$$\frac{24}{{}^-3-5i}\times\frac{{}^-3+5i}{{}^-3+5i}=\frac{{}^-72+120i}{9-25i^2}=\frac{{}^-72+120i}{9+25}=\frac{{}^-72+120i}{34}=\frac{{}^-36+60i}{17}$$

Divided everything by 2.

COMPETENCY 0005   **UNDERSTAND THE PRINCIPLES AND PROPERTIES OF ALGEBRAIC RELATIONS AND FUNCTIONS.**

- A **relation** is any set of ordered pairs.

- The **domain** of a relation is the set made of all the first coordinates of the ordered pairs.

- The **range** of a relation is the set made of all the second coordinates of the ordered pairs.

- A **function** is a relation in which different ordered pairs have different first coordinates. (No x values are repeated.)

- A **mapping** is a diagram with arrows drawn from each element of the domain to the corresponding elements of the range. If 2 arrows are drawn from the same element of the domain, then it is not a function.

- On a graph, use the **vertical line test** to look for a function. If any vertical line intersects the graph of a relation in more than one point, then the relation is not a function.

1. Determine the domain and range of this mapping.

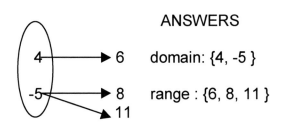

ANSWERS

domain: {4, -5 }

range : {6, 8, 11 }

2. Determine which of these are functions:

    a. $\{(1,{}^{-}4),(27,1)(94,5)(2,{}^{-}4)\}$

    b. $f(x) = 2x - 3$

    c. $A = \{(x,y) \mid xy = 24\}$

    d. $y = 3$

    e. $x = {}^{-}9$

    f. $\{(3,2),(7,7),(0,5),(2,{}^{-}4),(8,{}^{-}6),(1,0),(5,9),(6,{}^{-}4)\}$

3. Determine the domain and range of this graph.

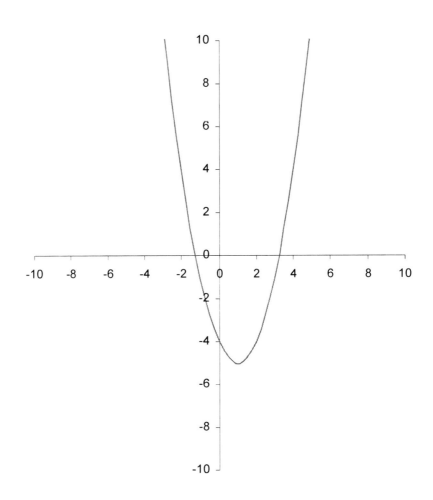

Language and Notation

A function can be defined as a set of ordered pairs in which each element of the domain is paired with one and only one element of the range. The symbol $f(x)$ is read "f of x." Letter other than "f" can be used to represent a function. The letter "g" is commonly used as in $g(x)$.

Sample problems:

1. Given $f(x) = 4x^2 - 2x + 3$, find $f(^-3)$.

(This question is asking for the range value that corresponds to the domain value of $^-3$).

$f(x) = 4x^2 - 2x + 3$	1. Replace x with $^-3$.
$f(^-3) = 4(^-3)^2 - 2(^-3) + 3$	
$f(^-3) = 45$	2. Solve.

2. Find f(3) and f(10), given $f(x) = 7$.

$f(x) = 7$	1. There are no x values
$(3) = 7$	to substitute for. This is
	your answer.
$f(x) = 7$	
$f10) = 7$	2. Same as above.

Notice that both answers are equal to the constant given.

Operations of Functions

If $f(x)$ is a function and the value of 3 is in the domain, the corresponding element in the range would be f(3). It is found by evaluating the function for $x = 3$. The same holds true for adding, subtracting, and multiplying in function form.

The symbol f^{-1} is read "the inverse of f". The $^-1$ is not an exponent. The inverse of a function can be found by reversing the order of coordinates in each ordered pair that satisfies the function. Finding the inverse functions means switching the place of x and y and then solving for y.

Sample problem:

1. Find $p(a+1)+3\{p(4a)\}$ if $p(x)=2x^2+x+1$.

Find $p(a+1)$.

$p(a+1)=2(a+1)^2+(a+1)+1$ Substitute $(a+1)$ for x.

$p(a+1)=2a^2+5a+4$ Solve.

Find $3\{p(4a)\}$.

$3\{p(4a)\}=3[2(4a)^2+(4a)+1]$ Substitute $(4a)$ for x, multiply by 3.

$3\{p(4a)\}=96a^2+12a+3$ Solve.

$p(a+1)+3\{p(4a)\}=2a^2+5a+4+96a^2+12a+3$

     Combine like terms.

$p(a+1)+3\{p(4a)\}=98a^2+17a+7$

Properties of Functions

Definition: A function f is even if $f(^-x)=f(x)$ and odd if $f(^-x)=^-f(x)$ for all x in the domain of f.

Sample problems:

Determine if the given function is even, odd, or neither even nor odd.

1. $f(x)=x^4-2x^2+7$

   $f(^-x)=(^-x)^4-2(^-x)^2+7$

   $f(^-x)=x^4-2x^2+7$

   $f(x)$ is an even function.

1. Find $f(^-x)$.
2. Replace x with ^-x.
3. Since $f(^-x)=f(x)$, $f(x)$ is an even function.

2. $f(x)=3x^3+2x$

   $f(^-x)=3(^-x)^3+2(^-x)$

   $f(^-x)=^-3x^3-2x$

1. Find $f(^-x)$.
2. Replace x with ^-x.
3. Since $f(x)$ is not equal to $f(^-x)$, $f(x)$ is not an even function.

   $^-f(x)=^-(3x^3+2x)$

   $^-f(x)=^-3x^3-2x$

   $f(x)$ is an odd function.

4. Try $^-f(x)$.
5. Since $f(^-x)=^-f(x)$, $f(x)$ is an odd function.

3. $g(x) = 2x^2 - x + 4$

   $g(^-x) = 2(^-x)^2 - (^-x) + 4$

   $g(^-x) = 2x^2 + x + 4$

   $^-g(x) = ^-(2x^2 - x + 4)$

   $^-g(x) = ^-2x^2 + x - 4$

   $g(x)$ is neither even nor odd.

1. First find $g(^-x)$.
2. Replace x with ^-x.
3. Since $g(x)$ does not equal $g(^-x)$, $g(x)$ is not an even function.
4. Try $^-g(x)$.
5. Since $^-g(x)$ does not equal $g(^-x)$, $g(x)$ is not an odd function.

There are 2 easy ways to find the **values of a function**. First to find the value of a function when $x = 3$, substitute 3 in place of every letter x. Then simplify the expression following the order of operations. For example, if $f(x) = x^3 - 6x + 4$, then to find f(3), substitute 3 for x.

The equation becomes $f(3) = 3^3 - 6(3) + 4 = 27 - 18 + 4 = 13$. So (3, 13) is a point on the graph of f(x).

A second way to find the value of a function is to use synthetic division. To find the value of a function when $x = 3$, divide 3 into the coefficients of the function. (Remember that coefficients of missing terms, like x^2, must be included). The remainder is the value of the function.

If $f(x) = x^3 - 6x + 4$, then to find f(3) using synthetic division:

Note the 0 for the missing x² term.

$$3 \overline{\left| \begin{array}{cccc} 1 & 0 & ^-6 & 4 \\ & 3 & 9 & 9 \\ \hline 1 & 3 & 3 & 13 \end{array} \right.}$$ ← this is the value of the function.

Therefore, (3, 13) is a point on the graph of $f(x) = x^3 - 6x + 4$.

Example: Find values of the function at integer values from $x = -3$ to $x = 3$ if $f(x) = x^3 - 6x + 4$.

If $x = {}^-3$:

$f({}^-3) = ({}^-3)^3 - 6({}^-3) + 4$

$\qquad = ({}^-27) - 6({}^-3) + 4$

$\qquad = {}^-27 + 18 + 4 = {}^-5$

synthetic division:

$$
{}^-3 \begin{array}{|rrrr} 1 & 0 & {}^-6 & 4 \\ & {}^-3 & 9 & {}^-9 \\ \hline \end{array}
$$

$\qquad 1 \; {}^-3 \quad 3 \; {}^-5 \leftarrow$ this is the value of the function if $x = {}^-3$.

$\qquad$ Therefore, $({}^-3, {}^-5)$ is a point on the graph.

If $x = {}^-2$:

$f({}^-2) = ({}^-2)^3 - 6({}^-2) + 4$

$\qquad = ({}^-8) - 6({}^-2) + 4$

$\qquad = {}^-8 + 12 + 4 = 8 \leftarrow$ this is the value of the function if $x = {}^-2$.

$\qquad$ Therefore, $({}^-2, 8)$ is a point on the graph.

If $x = {}^-1$:

$f({}^-1) = ({}^-1)^3 - 6({}^-1) + 4$

$\qquad = ({}^-1) - 6({}^-1) + 4$

$\qquad = {}^-1 + 6 + 4 = 9$

Synthetic division:

$$
{}^-1 \begin{array}{|rrrr} 1 & 0 & {}^-6 & 4 \\ & {}^-1 & 1 & 5 \\ \hline \end{array}
$$

$\qquad 1 \; {}^-1 \quad {}^-5 \quad 9 \leftarrow$ this is the value if the function if $x = {}^-1$.

$\qquad$ Therefore, $({}^-1, 9)$ is a point on the graph.

If $x = 0$:

$$f(0) = (0)^3 - 6(0) + 4$$
$$= 0 - 6(0) + 4$$
$$= 0 - 0 + 4 = 4 \leftarrow \text{this is the value of the function if } x = 0.$$

Therefore, $(0, 4)$ is a point on the graph.

If $x = 1$:

$$f(1) = (1)^3 - 6(1) + 4$$
$$= (1) - 6(1) + 4$$
$$= 1 - 6 + 4 = {}^-1$$

Synthetic division:

$$
1 \;\big|\;
\begin{array}{cccc}
1 & 0 & {}^-6 & 4 \\
  & 1 & 1 & {}^-5 \\
\hline
\end{array}
$$

$1 \quad 1 \quad {}^-5 \quad {}^-1 \leftarrow$ this is the value if the function of $x = 1$.

Therefore, $(1, {}^-1)$ is a point on the graph.

If $x = 2$:

$$f(2) = (2)^3 - 6(2) + 4$$
$$= 8 - 6(2) + 4$$
$$= 8 - 12 + 4 = 0$$

Synthetic division:

$$
2 \;\big|\;
\begin{array}{cccc}
1 & 0 & {}^-6 & 4 \\
  & 2 & 4 & {}^-4 \\
\hline
\end{array}
$$

$1 \quad 2 \quad {}^-2 \quad 0 \leftarrow$ this is the value of the function if $x = 2$.

Therefore, $(2, 0)$ is a point on the graph.

If $x = 3$:

$$f(3) = (3)^3 - 6(3) + 4$$
$$= 27 - 6(3) + 4$$
$$= 27 - 18 + 4 = 13$$

Synthetic division:

$$3 \overline{\begin{array}{cccc} 1 & 0 & ^-6 & 4 \\ & 3 & 9 & 9 \end{array}}$$
$$\quad 1 \quad 3 \quad 3 \quad 13 \leftarrow \text{this is the value of the function if } x = 3.$$

Therefore, (3, 13) is a point on the graph.

The following points are points on the graph:

X	Y
⁻3	⁻5
⁻2	8
⁻1	9
0	4
1	⁻1
2	0
3	13

Note the change in sign of the y value between $x = ^-3$ and $x = ^-2$. This indicates there is a zero between $x = ^-3$ and $x = ^-2$. Since there is another change in sign of the y value between $x = 0$ and $x = ^-1$, there is a second root there. When $x = 2$, $y = 0$ so $x = 2$ is an exact root of this polynomial.

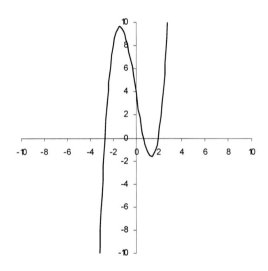

Domain and Range

- A **relation** is any set of ordered pairs.
- The **domain** of the relation is the set of all first co-ordinates of the ordered pairs. (These are the x coordinates.)
- The **range** of the relation is the set of all second co-ordinates of the ordered pairs. (These are the y coordinates.)

Practice Problems:

1. If $A = \left\{ (x,y) \mid y = x^2 - 6 \right\}$, find the domain and range.

2. Give the domain and range of set B if:

    $B = \left\{ (1, ^-2),(4, ^-2),(7, ^-2),(6, ^-2) \right\}$

3. Determine the domain of this function:

    $f(x) = \dfrac{5x + 7}{x^2 - 4}$

4. Determine the domain and range of these graphs.

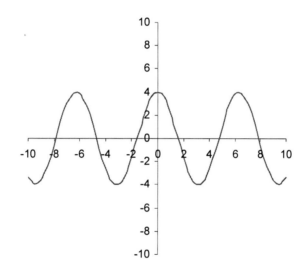

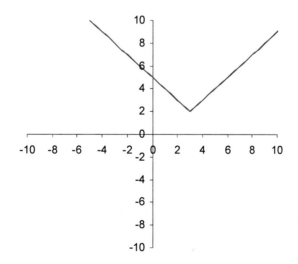

5. If $E = \left\{ (x,y) \mid y = 5 \right\}$, find the domain and range.

6. Determine the ordered pairs in the relation shown in this mapping.

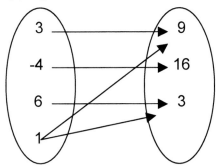

Different types of **function transformations** affect the graph and characteristics of a function in predictable ways. The basic types of transformation are horizontal and vertical shift (translation), horizontal and vertical scaling (dilation), and reflection. As an example of the types of transformations, we will consider transformations of the functions $f(x) = x^2$.

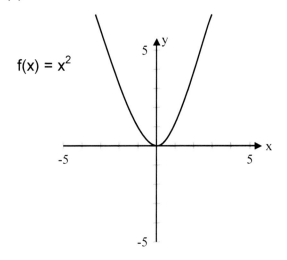

$f(x) = x^2$

Horizontal shifts take the form $g(x) = f(x \pm c)$. For example, we obtain the graph of the function $g(x) = (x + 2)^2$ by shifting the graph of $f(x) = x^2$ two units to the left. The graph of the function $h(x) = (x - 2)^2$ is the graph of $f(x) = x^2$ shifted two units to the right.

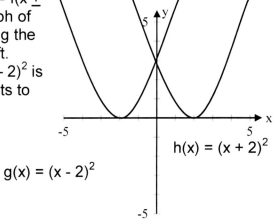

$h(x) = (x + 2)^2$

$g(x) = (x - 2)^2$

Vertical shifts take the form $g(x) = f(x) \pm c$. For example, we obtain the graph of the function $g(x) = (x^2) - 2$ by shifting the graph of $f(x) = x^2$ two units down. The graph of the function $h(x) = (x^2) + 2$ is the graph of $f(x) = x^2$ shifted two units up.

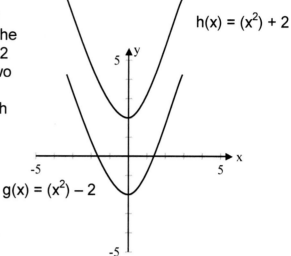

$h(x) = (x^2) + 2$

$g(x) = (x^2) - 2$

Horizontal scaling takes the form $g(x) = f(cx)$. For example, we obtain the graph of the function $g(x) = (2x)^2$ by compressing the graph of $f(x) = x^2$ in the x-direction by a factor of two. If $c > 1$ the graph is compressed in the x-direction, while if $1 > c > 0$ the graph is stretched in the x-direction.

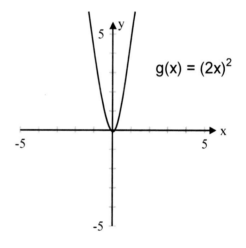

$g(x) = (2x)^2$

COMPETENCY 0006   **UNDERSTAND THE PRINCIPLES AND PROPERTIES OF LINEAR ALGEBRA.**

Addition of matrices is accomplished by adding the corresponding elements of the two matrices. Subtraction is defined as the inverse of addition. In other words, change the sign on all the elements in the second matrix and add the two matrices.

Sample problems:

Find the sum or difference.

1.
$$\begin{pmatrix} 2 & 3 \\ ^-4 & 7 \\ 8 & ^-1 \end{pmatrix} + \begin{pmatrix} 8 & ^-1 \\ 2 & ^-1 \\ 3 & ^-2 \end{pmatrix} =$$

$$\begin{pmatrix} 2+8 & 3+(^-1) \\ ^-4+2 & 7+(^-1) \\ 8+3 & ^-1+(^-2) \end{pmatrix}$$ Add corresponding elements.

$$\begin{pmatrix} 10 & 2 \\ ^-2 & 6 \\ 11 & ^-3 \end{pmatrix}$$ Simplify.

2.
$$\begin{pmatrix} 8 & ^-1 \\ 7 & 4 \end{pmatrix} - \begin{pmatrix} 3 & 6 \\ ^-5 & 1 \end{pmatrix} =$$

$$\begin{pmatrix} 8 & ^-1 \\ 7 & 4 \end{pmatrix} + \begin{pmatrix} ^-3 & ^-6 \\ 5 & ^-1 \end{pmatrix} =$$ Change all of the signs in the second matrix and then add the two matrices.

$$\begin{pmatrix} 8+(^-3) & ^-1+(^-6) \\ 7+5 & 4+(^-1) \end{pmatrix} =$$ Simplify.

$$\begin{pmatrix} 5 & ^-7 \\ 12 & 3 \end{pmatrix}$$

Practice problems:

1. $$\begin{pmatrix} 8 & ^-1 \\ 5 & 3 \end{pmatrix} + \begin{pmatrix} 3 & 8 \\ 6 & ^-2 \end{pmatrix} =$$ 2. $$\begin{pmatrix} 3 & 7 \\ ^-4 & 12 \\ 0 & ^-5 \end{pmatrix} - \begin{pmatrix} 3 & 4 \\ 6 & ^-1 \\ ^-5 & ^-5 \end{pmatrix} =$$

Scalar multiplication is the product of the scalar (the outside number) and each element inside the matrix.

Sample problem:

Given: $A = \begin{pmatrix} 4 & 0 \\ 3 & {}^{-}1 \end{pmatrix}$ Find 2A.

$$2A = 2\begin{pmatrix} 4 & 0 \\ 3 & {}^{-}1 \end{pmatrix}$$

$\begin{pmatrix} 2 \times 4 & 2 \times 0 \\ 2 \times 3 & 2 \times {}^{-}1 \end{pmatrix}$ Multiply each element in the

matrix by the scalar.

$\begin{pmatrix} 8 & 0 \\ 6 & {}^{-}2 \end{pmatrix}$ Simplify.

Practice problems:

1. $\quad {}^{-}2\begin{pmatrix} 2 & 0 & 1 \\ {}^{-}1 & {}^{-}2 & 4 \end{pmatrix}$ 2. $\quad 3\begin{pmatrix} 6 \\ 2 \\ 8 \end{pmatrix} + 4\begin{pmatrix} 0 \\ 7 \\ 2 \end{pmatrix}$

3. $\quad 2\begin{pmatrix} {}^{-}6 & 8 \\ {}^{-}2 & {}^{-}1 \\ 0 & 3 \end{pmatrix}$

The product of two matrices can only be found if the number of columns in the first matrix is equal to the number of rows in the second matrix. Matrix multiplication is not necessarily commutative.

Sample problems:

1. Find the product AB if:

$A = \begin{pmatrix} 2 & 3 & 0 \\ 1 & {}^{-}4 & {}^{-}2 \\ 0 & 1 & 1 \end{pmatrix}$ $B = \begin{pmatrix} {}^{-}2 & 3 \\ 6 & {}^{-}1 \\ 0 & 2 \end{pmatrix}$

$\qquad\qquad 3 \times 3 \qquad\qquad\qquad\qquad\qquad 3 \times 2$

Note: Since the number of columns in the first matrix ($3 \times \underline{3}$) matches the number of rows ($\underline{3} \times 2$) this product is defined and can be found. The dimensions of the product will be equal to the number of rows in the first matrix ($\underline{3} \times 3$) by the number of columns in the second matrix ($3 \times \underline{2}$). The answer will be a 3×2 matrix.

$$AB = \begin{pmatrix} 2 & 3 & 0 \\ 1 & ^-4 & ^-2 \\ 0 & 1 & 1 \end{pmatrix} \times \begin{pmatrix} ^-2 & 3 \\ 6 & ^-1 \\ 0 & 2 \end{pmatrix}$$

$$\begin{pmatrix} ^-2(^-2)+3(6)+0(0) & \\ & \\ & \end{pmatrix}$$ Multiply 1st row of A by 1st column of B.

$$\begin{pmatrix} 14 & 2(3)+3(^-1)+0(2) \\ & \\ & \end{pmatrix}$$ Multiply 1st row of A by 2nd column of B.

$$\begin{pmatrix} 14 & 3 \\ 1(^-2)-4(6)-2(0) & \\ & \end{pmatrix}$$ Multiply 2nd row of A by 1st column of B.

$$\begin{pmatrix} 14 & 3 \\ ^-26 & 1(3)-4(^-1)-2(2) \\ & \end{pmatrix}$$ Multiply 2nd row of A by 2nd column of B.

$$\begin{pmatrix} 14 & 3 \\ ^-26 & 3 \\ 0(^-2)+1(6)+1(0) & \end{pmatrix}$$ Multiply 3rd row of A by 1st column of B.

$$\begin{pmatrix} 14 & 3 \\ ^-26 & 3 \\ 6 & 0(3)+1(^-1)+1(2) \end{pmatrix}$$ Multiply 3rd row of A by 2nd column of B.

$$\begin{pmatrix} 14 & 3 \\ ^-26 & 3 \\ 6 & 1 \end{pmatrix}$$

The product of BA is not defined since the number of columns in B is not equal to the number of rows in A.

Practice problems:

1. $\begin{pmatrix} 3 & 4 \\ {}^-2 & 1 \end{pmatrix}\begin{pmatrix} {}^-1 & 7 \\ {}^-3 & 1 \end{pmatrix}$

2. $\begin{pmatrix} 1 & {}^-2 \\ 3 & 4 \\ 2 & 5 \\ -1 & 6 \end{pmatrix}\begin{pmatrix} 3 & {}^-1 & {}^-4 \\ {}^-1 & 2 & 3 \end{pmatrix}$

A **matrix** is a square array of numbers called its entries or elements. The dimensions of a matrix are written as the number of rows (r) by the number of columns (r × c).

$\begin{pmatrix} 1 & 2 & 3 \\ 4 & 5 & 6 \end{pmatrix}$ is a 2 × 3 matrix (2 rows by 3 columns)

$\begin{pmatrix} 1 & 2 \\ 3 & 4 \\ 5 & 6 \end{pmatrix}$ is a 3 × 2 matrix (3 rows by 2 columns)

Associated with every square matrix is a number called the determinant. Use these formulas to calculate determinants.

2×2 $\begin{pmatrix} a & b \\ c & d \end{pmatrix} = ad - bc$

3×3

$\begin{pmatrix} a_1 & b_1 & c_1 \\ a_2 & b_2 & c_2 \\ a_3 & b_3 & c_3 \end{pmatrix} = (a_1b_2c_3 + b_1c_2a_3 + c_1a_2b_3) - (a_3b_2c_1 + b_3c_2a_1 + c_3a_2b_1)$

This is found by repeating the first two columns and then using the diagonal lines to find the value of each expression as shown below:

$$\begin{pmatrix} a_1^* & b_1^\circ & c_1^\bullet \\ a_2 & b_2^* & c_2^\circ \\ a_3 & b_3 & c_3^* \end{pmatrix}\begin{matrix} a_1 & b_1 \\ a_2^\bullet & b_2 \\ a_3^\circ & b_3^\bullet \end{matrix}$$
$$= (a_1b_2c_3 + b_1c_2a_3 + c_1a_2b_3) - (a_3b_2c_1 + b_3c_2a_1 + c_3a_2b_1)$$

Sample Problem:

1. Find the value of the determinant:

$$\begin{pmatrix} 4 & {}^{-}8 \\ 7 & 3 \end{pmatrix} = (4)(3) - (7)({}^{-}8) \quad \text{Cross multiply and subtract.}$$

$$12 - ({}^{-}56) = 68 \qquad \text{Then simplify.}$$

When given the following **system of equations**:

$$ax + by = e$$
$$cx + dy = f$$

the matrix equation is written in the form:

$$\begin{pmatrix} a & b \\ c & d \end{pmatrix}\begin{pmatrix} x \\ y \end{pmatrix} = \begin{pmatrix} e \\ f \end{pmatrix}$$

The solution is found using the inverse of the matrix of coefficients. Inverse of matrices can be written as follows:

$$A^{-1} = \frac{1}{\text{determinant of } A}\begin{pmatrix} d & {}^{-}b \\ {}^{-}c & a \end{pmatrix}$$

Sample Problem:
1. Write the matrix equation of the system.

$$3x - 4y = 2$$
$$2x + y = 5$$

$$\begin{pmatrix} 3 & {}^{-}4 \\ 2 & 1 \end{pmatrix}\begin{pmatrix} x \\ y \end{pmatrix} = \begin{pmatrix} 2 \\ 5 \end{pmatrix} \qquad \text{Definition of matrix equation.}$$

$$\begin{pmatrix} x \\ y \end{pmatrix} = \frac{1}{11}\begin{pmatrix} 1 & 4 \\ {}^{-}2 & 3 \end{pmatrix}\begin{pmatrix} 2 \\ 5 \end{pmatrix} \qquad \text{Multiply by the inverse of the coefficient matrix.}$$

$$\begin{pmatrix} x \\ y \end{pmatrix} = \frac{1}{11}\begin{pmatrix} 22 \\ 11 \end{pmatrix} \qquad \text{Matrix multiplication.}$$

$$\begin{pmatrix} x \\ y \end{pmatrix} = \begin{pmatrix} 2 \\ 1 \end{pmatrix} \qquad \text{Scalar multiplication. The solution is (2,1).}$$

Practice problems:

1. $\begin{aligned} x + 2y &= 5 \\ 3x + 5y &= 14 \end{aligned}$

2. $\begin{aligned} {}^-3x + 4y - z &= 3 \\ x + 2y - 3z &= 9 \\ y - 5z &= {}^-1 \end{aligned}$

The variable in a **matrix equation** represents a matrix. When solving for the answer use the adding, subtracting and scalar multiplication properties.

Sample problem:

Solve the matrix equation for the variable X.

$$2x + \begin{pmatrix} 4 & 8 & 2 \\ 7 & 3 & 4 \end{pmatrix} = 2\begin{pmatrix} 1 & {}^-2 & 0 \\ 3 & {}^-5 & 7 \end{pmatrix}$$

$$2x = 2\begin{pmatrix} 1 & {}^-2 & 0 \\ 3 & {}^-5 & 7 \end{pmatrix} - \begin{pmatrix} 4 & 8 & 2 \\ 7 & 3 & 4 \end{pmatrix}$$
Subtract $\begin{pmatrix} 4 & 8 & 2 \\ 7 & 3 & 4 \end{pmatrix}$ from both sides.

$$2x = \begin{pmatrix} 2 & {}^-4 & 0 \\ 6 & {}^-10 & 14 \end{pmatrix} + \begin{pmatrix} {}^-4 & {}^-8 & {}^-2 \\ {}^-7 & {}^-3 & {}^-4 \end{pmatrix}$$
Scalar multiplication and matrix subtraction.

$$2x = \begin{pmatrix} {}^-2 & {}^-12 & {}^-2 \\ {}^-1 & {}^-13 & 10 \end{pmatrix}$$
Matrix addition.

$$x = \begin{pmatrix} {}^-1 & {}^-6 & {}^-1 \\ -\dfrac{1}{2} & -\dfrac{13}{2} & 5 \end{pmatrix}$$
Multiply both sides by $\dfrac{1}{2}$.

Solve for the unknown values of the elements in the matrix.

$$\begin{pmatrix} x+3 & y-2 \\ z+3 & w-4 \end{pmatrix} + \begin{pmatrix} {}^-2 & 4 \\ 2 & 5 \end{pmatrix} = \begin{pmatrix} 4 & 8 \\ 6 & 1 \end{pmatrix}$$

$$\begin{pmatrix} x+1 & y+2 \\ z+5 & w+1 \end{pmatrix} = \begin{pmatrix} 4 & 8 \\ 6 & 1 \end{pmatrix}$$
Matrix addition.

$\begin{aligned} x+1 &= 4 & y+2 &= 8 & z+5 &= 6 & w+1 &= 1 \\ x &= 3 & y &= 6 & z &= 1 & w &= 0 \end{aligned}$
Definition of equal matrices.

Practice problems:

1. $x + \begin{pmatrix} 7 & 8 \\ 3 & ^{-}1 \\ 2 & ^{-}3 \end{pmatrix} = \begin{pmatrix} 0 & 8 \\ ^{-}9 & ^{-}4 \\ 8 & 2 \end{pmatrix}$

2. $4x - 2\begin{pmatrix} 0 & 10 \\ 6 & ^{-}4 \end{pmatrix} = 3\begin{pmatrix} 4 & 9 \\ 0 & 12 \end{pmatrix}$

3. $\begin{pmatrix} 7 & 3 \\ 2 & 4 \\ 3 & 7 \end{pmatrix} + \begin{pmatrix} a+2 & b+4 \\ c-3 & d+1 \\ e & f+3 \end{pmatrix} = \begin{pmatrix} 4 & 6 \\ ^{-}1 & 1 \\ 3 & 0 \end{pmatrix}$

A **transformation matrix** defines how to map points from one coordinate space into another coordinate space. The matrix used to accomplish two-dimensional transformations is described mathematically by a 3-by-3 matrix.

Example:

A point transformed by a 3-by-3 matrix

$$[x \quad y \quad 1] \quad x \quad \begin{pmatrix} a & b & u \\ c & d & v \\ t_x & t_y & w \end{pmatrix} = [x' \quad y' \quad 1]$$

A 3-by-3 matrix transforms a point (x, y) into a point (x', y') by means of the following equations:

$$x' = ax + cy + t_x$$

$$y' = bx + dy + t_y$$

COMPETENCY 0007   **UNDERSTAND THE PROPERTIES OF LINEAR FUNCTIONS AND RELATIONS.**

- A first degree equation has an equation of the form $ax + by = c$. To find the slope of a line, solve the equation for y. This gets the equation into **slope intercept form**, $y = mx + b$. The value m is the line's slope.

- To find the y intercept, substitute 0 for x and solve for y. This is the y intercept. The y intercept is also the value of b in $y = mx + b$.

- To find the x intercept, substitute 0 for y and solve for x. This is the x intercept.

- If the equation solves to **x = any number**, then the graph is a **vertical line**. It only has an x intercept. Its slope is **undefined**.

- If the equation solves to **y = any number**, then the graph is a **horizontal line**. It only has a y intercept. Its slope is 0 (zero).

1. Find the slope and intercepts of $3x + 2y = 14$.

$$3x + 2y = 14$$
$$2y = {}^-3x + 14$$
$$y = {}^-3/2\ x + 7$$

The slope of the line is $^-3/2$, the value of m.
The y intercept of the line is 7.

The intercepts can also be found by substituting 0 in place of the other variable in the equation.

To find the y intercept:
let $x = 0$; $3(0) + 2y = 14$
$0 + 2y = 14$
$2y = 14$
$y = 7$
$(0,7)$ is the y intercept.

To find the x intercept:
let $y = 0$; $3x + 2(0) = 14$
$3x + 0 = 14$
$3x = 14$
$x = 14/3$
$(14/3, 0)$ is the x intercept.

Find the slope and the intercepts (if they exist) for these equations:

1. $5x + 7y = {}^-70$
2. $x - 2y = 14$
3. $5x + 3y = 3(5 + y)$
4. $2x + 5y = 15$

From a Graph

The equation of a graph can be found by finding its slope and its y-intercept. To find the slope, find two points on the graph where co-ordinates are integer values. Using points: (x_1, y_1) and (x_2, y_2).

$$\text{slope} = \frac{y_2 - y_1}{x_2 - x_1}$$

The y-intercept is the y-coordinate of the point where a line crosses the y-axis. The equation can be written in slope-intercept form, which is $y = mx + b$, where m is the slope and b is the y-intercept. To rewrite the equation into some other form, multiply each term by the common denominator of all the fractions. Then rearrange terms as necessary.

If the graph is a **vertical line**, then the equation solves to **x = the x-coordinate of any point on the line**.

If the graph is a **horizontal line**, then the equation solves to **y = the y-coordinate of any point on the line**.

From Two Points

Given two points on a line, the first thing to do is to find the slope of the line. If two points on the graph are (x_1, y_1) and (x_2, y_2), then the slope is found using the formula:

$$\text{slope} = \frac{y_2 - y_1}{x_2 - x_1}$$

The slope will now be denoted by the letter **m**. To write the equation of a line, choose either point. Substitute them into the formula:

$$Y - y_a = m(X - x_a)$$

Remember (x_a, y_a) can be (x_1, y_1) or (x_2, y_2). If **m**, the value of the slope, is distributed through the parentheses, the equation can be rewritten into other forms of the equation of a line.

Find the equation of a line through $(9, -6)$ and $(-1, 2)$.

$$\text{slope} = \frac{y_2 - y_1}{x_2 - x_1} = \frac{2 - (-6)}{-1 - 9} = \frac{8}{-10} = -\frac{4}{5}$$

$$Y - y_a = m(X - x_a) \rightarrow Y - 2 = -4/5(X - (-1)) \rightarrow$$
$$Y - 2 = -4/5(X + 1) \rightarrow Y - 2 = -4/5\,X - 4/5 \rightarrow$$
$$Y = -4/5\,X + 6/5 \quad \text{This is the slope-intercept form.}$$

Multiplying by 5 to eliminate fractions, it is:

$$5Y = -4X + 6 \rightarrow 4X + 5Y = 6 \quad \text{Standard form.}$$

Write the equation of a line through these two points:

1. $(5, 8)$ and $(-3, 2)$
2. $(11, 10)$ and $(11, -3)$
3. $(-4, 6)$ and $(6, 12)$
4. $(7, 5)$ and $(-3, 5)$

Graphing Linear Equations and Inequalities

A first degree equation has an equation of the form $ax + by = c$. To graph this equation, find both one point and the slope of the line or find two points. To find a point and slope, solve the equation for y. This gets the equation in **slope intercept form**, $y = mx + b$. The point (0, b) is the y-intercept and m is the line's slope. To find any two points, substitute any two numbers for x and solve for y. To find the intercepts, substitute 0 for x and then 0 for y.

Remember that graphs will go up as they go to the right when the slope is positive. Negative slopes make the lines go down as they go to the right.

If the equation solves to *x* = **any number**, then the graph is a **vertical line**. It only has an *x* intercept. Its slope is **undefined**.

If the equation solves to *y* = **any number**, then the graph is a **horizontal line**. It only has a *y* intercept. Its slope is 0 (zero).

When graphing a linear inequality, the line will be dotted if the inequality sign is < or >. If the inequality signs are either ≥ or ≤ , the line on the graph will be a solid line. Shade above the line when the inequality sign is ≥ or >. Shade below the line when the inequality sign is < or≤. Inequalities of the form *x* >, *x* ≤, *x* <, or *x* ≥ number, draw a vertical line (solid or dotted). Shade to the right for > or≥. Shade to the left for < or≤. Remember: **Dividing or multiplying by a negative number will reverse the direction of the inequality sign.**

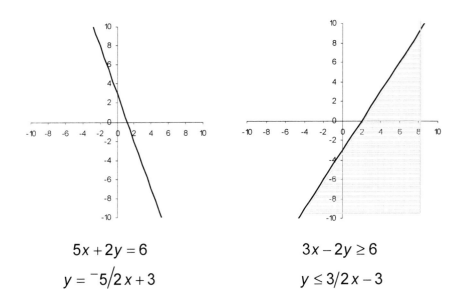

$$5x + 2y = 6$$

$$y = {}^-5/2\,x + 3$$

$$3x - 2y \geq 6$$

$$y \leq 3/2\,x - 3$$

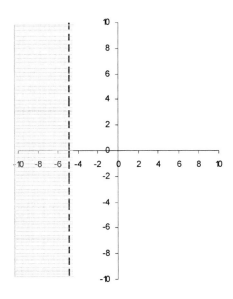

$$3x + 12 < -3$$
$$x < {}^-5$$

Graph the following:

1. $2x - y = {}^-4$
2. $x + 3y > 6$
3. $3x + 2y \leq 2y - 6$

Algebraic and Graphical Problem Solving

Word problems can sometimes be solved by using a system of two equations in two unknowns. This system can then be solved using substitution, the addition-subtraction method, or graphing.

Example: Mrs. Winters bought 4 dresses and 6 pairs of shoes for $340. Mrs. Summers went to the same store and bought 3 dresses and 8 pairs of shoes for $360. If all the dresses were the same price and all the shoes were the same price, find the price charged for a dress and for a pair of shoes.

Let x = price of a dress
Let y = price of a pair of shoes

Then Mrs. Winters' equation would be: $4x + 6y = 340$
Mrs. Summers' equation would be: $3x + 8y = 360$

To solve by addition-subtraction:

Multiply the first equation by 4: $4(4x + 6y = 340)$
Multiply the other equation by -3: $-3(3x + 8y = 360)$
By doing this, the equations can be added to each other to eliminate one variable and solve for the other variable.

$$16x + 24y = 1360$$
$$\underline{-9x - 24y = -1080}$$
$$7x = 280$$
$$x = 40 \leftarrow \text{the price of a dress was } \$40$$

solving for y, $y = 30$ $\leftarrow$ the price of a pair of shoes, $30

Example: Aardvark Taxi charges $4 initially plus $1 for every mile traveled. Baboon Taxi charges $6 initially plus $.75 for every mile traveled. Determine when it is cheaper to ride with Aardvark Taxi or to ride with Baboon Taxi.

Aardvark Taxi's equation:	$y = 1x + 4$
Baboon Taxi's equation :	$y = .75x + 6$
Using substitution:	$.75x + 6 = x + 4$
Multiplying by 4:	$3x + 24 = 4x + 16$
Solving for x :	$8 = x$

This tells you that at 8 miles the total charge for the two companies is the same. If you compare the charge for 1 mile, Aardvark charges $5 and Baboon charges $6.75. Clearly Aardvark is cheaper for distances up to 8 miles, but Baboon Taxi is cheaper for distances greater than 8 miles.

This problem can also be solved by graphing the 2 equations.

$$y = 1x + 4 \qquad y = .75x + 6$$

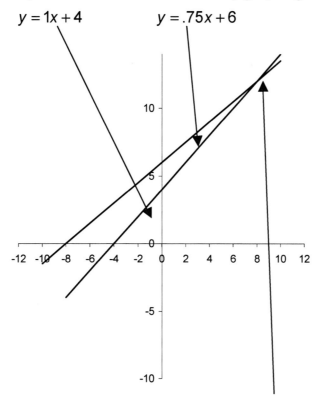

The lines intersect at (8, 12), therefore at 8 miles, both companies charge $12. At values less than 8 miles, Aardvark Taxi charges less (the graph is below Baboon). Greater than 8 miles, Aardvark charges more (the graph is above Baboon).

Example: The YMCA wants to sell raffle tickets to raise at least $32,000. If they must pay $7,250 in expenses and prizes out of the money collected from the tickets, how many tickets worth $25 each must they sell?

Solution: Since they want to raise **at least $32,000**, that means they would be happy to get 32,000 **or more**. This requires an inequality.

Let x = number of tickets sold
Then $25x$ = total money collected for x tickets

Total money minus expenses is greater than $32,000.

$$25x - 7250 \geq 32000$$
$$25x \geq 39250$$
$$x \geq 1570$$

If they sell **1,570 tickets or more**, they will raise AT LEAST $32,000.

Example: The Simpsons went out for dinner. All four of them ordered the aardvark steak dinner. Bert paid for the 4 meals and included a tip of $12 for a total of $84.60. How much was an aardvark steak dinner?

Let $x =$ the price of one aardvark dinner.
So $4x =$ the price of 4 aardvark dinners.

$$4x + 12 = 84.60$$
$$4x = 72.60 \qquad \text{[SA6]}$$
$$x = \$18.15 \text{ for each dinner.}$$

Word problems can sometimes be solved by using a system of two equations in 2 unknowns. This system can then be solved using **substitution**, or the **addition-subtraction method**.

Example: Farmer Greenjeans bought 4 cows and 6 sheep for $1700. Mr. Ziffel bought 3 cows and 12 sheep for $2400. If all the cows were the same price and all the sheep were another price, find the price charged for a cow or for a sheep.

Let x = price of a cow
Let y = price of a sheep

Then Farmer Greenjeans' equation would be: $4x + 6y = 1700$
Mr. Ziffel's equation would be: $3x + 12y = 2400$

To solve by **addition-subtraction**:

Multiply the first equation by -2 [SA7]: $-2(4x + 6y = 1700)$ [SA8]
Keep the other equation the same: $(3x + 12y = 2400)$

By doing this, the equations can be added to each other to eliminate one variable and solve for the other variable.

$$-8x - 12y = -3400$$
$$\underline{3x + 12y = 2400} \qquad \text{Add these equations.} \text{[SA9]}$$
$$-5x \qquad = -1000$$

$x = 200 \leftarrow$ the price of a cow was \$200.
Solving for y, $y = 150 \leftarrow$ the price of a sheep was \$150.

(This problem can also be solved by substitution or determinants.)

<u>Example</u>: John has 9 coins, which are either dimes or nickels, that are worth \$.65. Determine how many of each coin he has.

Let d = number of dimes.
Let n = number of nickels.

The number of coins total 9.
The value of the coins equals 65.

Then: $n + d = 9$
$5n + 10d = 65$

Multiplying the first equation by $^-5$, it becomes:

$$^-5n - 5d = {}^-45$$
$$\underline{5n + 10d = 65}$$
$$5d = 20$$

$d = 4$ There are 4 dimes, so there are
$(9 - 4)$ or 5 nickels.

<u>Example</u>: Sharon's Bike Shoppe can assemble a 3 speed bike in 30 minutes or a 10 speed bike in 60 minutes. The profit on each bike sold is \$60 for a 3 speed or \$75 for a 10 speed bike. How many of each type of bike should they assemble during an 8 hour day (480 minutes) to make the maximum profit? Total daily profit must be at least \$300.

Let x = number of 3 speed bikes.
y = number of 10 speed bikes.

Since there are only 480 minutes to use each day,

$30x + 60y \le 480$ is the first inequality.

Since the total daily profit must be at least \$300,

$60x + 75y \geq 300$ is the second inequality.

$30x + 60y \leq 480$ solves to $y \leq 8 - 1/2x$
$60y \leq -30x + 480$

$$y \leq -\frac{1}{2}x + 8$$

$60x + 75y \geq 300$ solves to $y \geq 4 - 4/5x$
$75y + 60x \geq 300$

$$75y \geq -60x + 300$$

$$y \geq -\frac{4}{5}x + 4$$

Graph these 2 inequalities:
$y \leq 8 - 1/2x$
$y \geq 4 - 4/5x$

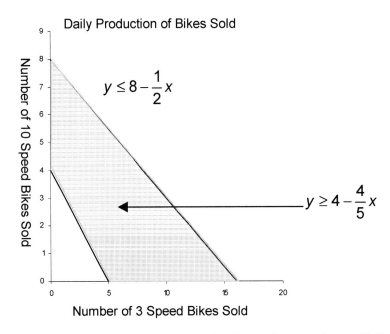

Realize that $x \geq 0$ and $y \geq 0$, since the number of bikes assembled can not be a negative number. Graph these as additional constraints on the problem. The number of bikes assembled must always be an integer value, so points within the shaded area of the graph must have integer values. The maximum profit will occur at or near a corner of the shaded portion of this graph. Those points occur at (0,4), (0,8), (16,0), or (5,0).

Since profits are $\$60/3$-speed or $\$75/10$-speed, the profit would be :

$$(0,4) \quad 60(0) + 75(4) = 300$$
$$(0,8) \quad 60(0) + 75(8) = 600$$
$$(16,0) \quad 60(16) + 75(0) = 960 \leftarrow \text{Maximum profit}$$
$$(5,0) \quad 60(5) + 75(0) = 300$$

The maximum profit would occur if 16 3-speed bikes are made daily.

To solve an **equation or inequality**, follow these steps:

STEP   1.   If there are parentheses, use the distributive property to eliminate them.

STEP   2.   If there are fractions, determine their LCD (least common denominator). Multiply every term of the equation by the LCD. This will cancel out all of the fractions while solving the equation or inequality.

STEP   3. If there are decimals, find the largest decimal. Multiply each term by a power of 10(10, 100, 1000,etc.) with the same number of zeros as the length of the decimal. This will eliminate all decimals while solving the equation or inequality.

STEP   4.   Combine like terms on each side of the equation or inequality.

STEP   5. If there are variables on both sides of the equation, add or subtract one of those variable terms to move it to the other side. Combine like terms.

STEP   6.   If there are constants on both sides, add or subtract one of those constants to move it to the other side. Combine like terms.

STEP   7. If there is a coefficient in front of the variable, divide both sides by this number. This is the answer to an equation. However, remember:

           Dividing or multiplying an inequality by a negative number will reverse the direction of the inequality sign.

STEP   8. The solution of a linear equation solves to one single number. The solution of an inequality is always stated including the inequality sign.

<u>Example:</u>   Solve: $3(2x+5)-4x=5(x+9)$

$6x+15-4x=5x+45$ ref. step 1

$2x+15=5x+45$ ref. step 4

$^-3x+15=45$ ref. step 5

$^-3x=30$ ref. step 6

$x=^-10$ ref. step 7

<u>Example:</u>   Solve: $1/2(5x+34)=1/4(3x-5)$

$5/2x+17=3/4x-5/4$ ref.step 1

LCD of 5/2, 3/4, and 5/4 is 4.

Multiply by the LCD of 4.

$4(5/2x+17)=(3/4x-5/4)4$ ref.step 2

$10x+68=3x-5$

$7x+68=^-5$ ref.step 5

$7x=^-73$ ref.step 6

$x=^-73\!\!\Big/\!\!_7$ or $^-10\,^3\!\!\Big/\!\!_7$ ref.step 7

Check:

$$\frac{1}{2}\left[5\frac{-13}{7}+34\right]=\frac{1}{4}\left[3\left(\frac{-13}{7}\right)-\frac{5}{4}\right]$$

$$\frac{1}{2}\left[\frac{-13(5)}{7}+34\right]=\frac{1}{4}\left[3\left(\frac{-13}{7}\right)-\frac{5}{4}\right]$$

$$\frac{-13(5)}{7}+17=\frac{3(-13)}{28}-\frac{5}{4}$$

$$\frac{-13(5)+17(14)}{14}=\frac{3(-13)}{28}-\frac{5}{4}$$

$$\left[-13(5)+17(14)\right]2=\frac{3(-13)-35}{28}$$

$$\frac{-130+476}{28}=\frac{-219-35}{28}$$

$$\frac{-254}{28}=\frac{-254}{28}$$

<u>Example:</u>   Solve: $6x + 21 < 8x + 31$

$$^-2x + 21 < 31 \qquad \text{ref. step 5}$$
$$^-2x < 10 \qquad \text{ref. step 6}$$
$$x > {}^-5 \qquad \text{ref. step 7}$$

Note that the inequality sign has changed.

Word Problems

To solve by **substitution**:

Solve one of the equations for a variable. (Try to make an equation without fractions if possible.) Substitute this expression into the equation that you have not yet used. Solve the resulting equation for the value of the remaining variable.

$$4x + 6y = 1700$$
$$3x + 12y = 2400 \leftarrow \text{Solve this equation for } x.$$

It becomes $x = 800 - 4y$. Now substitute $800 - 4y$ in place of x in the OTHER equation. $4x + 6y = 1700$ now becomes:

$$4(800 - 4y) + 6y = 1700$$
$$3200 - 16y + 6y = 1700$$
$$3200 - 10y = 1700 \qquad \text{[SA10]}$$
$$-10y = -1500$$
$$y = 150, \text{ or \$150 for a sheep.}$$

Substituting 150 back into an equation for y, find x.
$$4x + 6(150) = 1700$$
$$4x + 900 = 1700$$
$$4x = 800 \text{ so } x = 200 \text{ for a cow.}$$

To solve by **determinants**:

Let x = price of a cow
Let y = price of a sheep
Then Farmer Greenjeans' equation would be: $4x + 6y = 1700$
Mr. Ziffel's equation would be: $3x + 12y = 2400$

To solve this system using determinants, make one 2 by 2 determinant divided by another 2 by 2 determinant. The bottom determinant is filled with the x and y term coefficients. The top determinant is almost the same as this bottom determinant. The only difference is that when you are solving for x, the x coefficients are replaced with the constants. Likewise, when you are solving for y, the y coefficients are replaced with the constants. To find the value of a 2 by 2 determinant, $\begin{pmatrix} a & b \\ c & d \end{pmatrix}$, is found by $ad - bc$.

$$x = \frac{\begin{pmatrix} 1700 & 6 \\ 2400 & 12 \end{pmatrix}}{\begin{pmatrix} 4 & 6 \\ 3 & 12 \end{pmatrix}} = \frac{1700(12) - 6(2400)}{4(12) - 6(3)} = \frac{20400 - 14400}{48 - 18} = \frac{6000}{30} = 200$$

$$y = \frac{\begin{pmatrix} 4 & 1700 \\ 3 & 2400 \end{pmatrix}}{\begin{pmatrix} 4 & 6 \\ 3 & 12 \end{pmatrix}} = \frac{2400(4) - 3(1700)}{4(12) - 6(3)} = \frac{9600 - 5100}{48 - 18} = \frac{4500}{30} = 150$$

NOTE: The bottom determinant is always the same value for each letter.

Word problems can sometimes be solved by using a system of three equations in 3 unknowns. This system can then be solved using **substitution**, the **addition-subtraction method**, or **determinants**.

Example: Mrs. Allison bought 1 pound of potato chips, a 2-pound beef roast, and 3 pounds of apples for a total of $8.19. Mr. Bromberg bought a 3-pound beef roast and 2 pounds of apples for $9.05. Kathleen Kaufman bought 2 pounds of potato chips, a 3-pound beef roast, and 5 pounds of apples for $13.25. Find the per pound price of each item.

Let x = price of a pound of potato chips
Let y = price of a pound of roast beef
Let z = price of a pound of apples

Mrs. Allison's equation would be: $\quad 1x + 2y + 3z = 8.19$
Mr. Bromberg's equation would be: $\quad 3y + 2z = 9.05$
K. Kaufman's equation would be: $\quad 2x + 3y + 5z = 13.25$

Take the first equation and solve it for x. (This was chosen because x is the easiest variable to get alone in this set of equations.) This equation would become:

$$x = 8.19 - 2y - 3z$$

Substitute this expression into the other equations in place of the letter x:

$$3y + 2z = 9.05 \leftarrow \text{equation 2}$$
$$2(8.19 - 2y - 3z) + 3y + 5z = 13.25 \leftarrow \text{equation 3}$$

Simplify the equation by combining like terms:

$$3y + 2z = 9.05 \leftarrow \text{equation 2}$$
$$* \quad -1y - 1z = -3.13 \leftarrow \text{equation 3}$$

Solve equation 3 for either y or z:

$$y = 3.13 - z \quad \text{Substitute this into equation 2 for } y:$$

$$3(3.13 - z) + 2z = 9.05 \leftarrow \text{equation 2}$$
$$-1y - 1z = -3.13 \leftarrow \text{equation 3}$$

Combine like terms in equation 2:
$$9.39 - 3z + 2z = 9.05$$
$$z = .34 \quad \text{per pound price of apples}$$

Substitute .34 for z in the starred equation above to solve for y:
$$y = 3.13 - z \text{ becomes } y = 3.13 - .34, \text{ so}$$
$$y = 2.79 = \text{per pound price of roast beef}$$

Substituting .34 for z and 2.79 for y in one of the original equations, solve for x:

$$1x + 2y + 3z = 8.19$$
$$1x + 2(2.79) + 3(.34) = 8.19$$
$$x + 5.58 + 1.02 = 8.19$$
$$x + 6.60 = 8.19$$
$$x = 1.59 \quad \text{per pound of potato chips}$$

$$(x, y, z) = (1.59, 2.79, .34)$$

To solve by **addition-subtraction**:

Choose a letter to eliminate. Since the second equation is already missing an x, eliminate x from equations 1 and 3.

1) $1x + 2y + 3x = 8.19 \leftarrow$ Multiply by -2 below.
2) $3y + 2z = 9.05$
3) $2x + 3y + 5z = 13.25$

$-2(1x + 2y + 3z = 8.19) = -2x - 4y - 6z = -16.38$ [SA11]
Keep equation 3 the same : $2x + 3y + 5z = 13.25$
By doing this, the equations
can be added to each other to
eliminate one variable.

$-y - z = -3.13 \leftarrow$ equation 4 [SA12]

The equations left to solve are equations 2 and 4:
   $-y - z = -3.13 \leftarrow$ equation 4
   $3y + 2z = 9.05 \leftarrow$ equation 2

Multiply equation 4 by 3: $3(-y - z = -3.13)$
Keep equation 2 the same: $3y + 2z = 9.05$

$$-3y - 3z = -9.39$$
$$\underline{3y + 2z = 9.05}$$ Add these equations. [SA13]
$$-1z = -.34$$
$z = .34 \leftarrow$ the per pound price of apples
solving for y, $y = 2.79 \leftarrow$ the per pound roast beef price
solving for x, $x = 1.59 \leftarrow$ potato chips, per pound price

To solve by **substitution**:

Solve one of the 3 equations for a variable. (Try to make an equation without fractions if possible.) Substitute this expression into the other 2 equations that you have not yet used.

1) $1x + 2y + 3z = 8.19 \leftarrow$ Solve for x.
2) $3y + 2z = 9.05$
3) $2x + 3y + 5z = 13.25$
   Equation 1 becomes $x = 8.19 - 2y - 3z$.

Substituting this into equations 2 and 3, they become:

2) $3y + 2z = 9.05$

3) $2(8.19 - 2y - 3z) + 3y + 5z = 13.25$

   $16.38 - 4y - 6z + 3y + 5z = 13.25$

   $^-y - z = {}^-3.13$

The equations left to solve are :

   $3y + 2z = 9.05$

   $^-y - z = {}^-3.13 \leftarrow$ Solve for either y or z.

It becomes $y = 3.13 - z$. Now substitute $3.13 - z$ in place of y in the OTHER equation. $3y + 2z = 9.05$ now becomes:

$$3(3.13 - z) + 2z = 9.05$$
$$9.39 - 3z + 2z = 9.05$$
$$9.39 - z = 9.05$$
$$^-z = {}^-.34$$
$$z = .34, \text{ or } \$.34/\text{lb of apples}$$

Substituting .34 back into an equation for z, find y.

   $3y + 2z = 9.05$

   $3y + 2(.34) = 9.05$

   $3y + .68 = 9.05$ so $y = 2.79/\text{lb of roast beef}$

Substituting .34 for z and 2.79 for y into one of the original equations, it becomes:

   $2x + 3y + 5z = 13.25$

   $2x + 3(2.79) + 5(.34) = 13.25$

   $2x + 8.37 + 1.70 = 13.25$

   $2x + 10.07 = 13.25$, so $x = 1.59/\text{lb of potato chips}$

To solve by **determinants**:

   Let x = price of a pound of potato chips
   Let y = price of a pound of roast beef
   Let z = price of a pound of apples

1) $1x + 2y + 3z = 8.19$

2) $3y + 2z = 9.05$

3) $2x + 3y + 5z = 13.25$

To solve this system using determinants, make one 3 by 3 determinant divided by another 3 by 3 determinant. The bottom determinant is filled with the x, y, and z term coefficients. The top determinant is almost the same as this bottom determinant. The only difference is that when you are solving for x, the x coefficients are replaced with the constants. When you are solving for y, the y coefficients are replaced with the constants. Likewise, when you are solving for z, the z coefficients are replaced with the constants. To find the value of a 3 by 3 determinant,

$$\begin{pmatrix} a & b & c \\ d & e & f \\ g & h & i \end{pmatrix} \text{ is found by the following steps:}$$

Copy the first two columns to the right of the determinant:

$$\begin{pmatrix} a & b & c \\ d & e & f \\ g & h & i \end{pmatrix} \begin{matrix} a & b \\ d & e \\ g & h \end{matrix}$$

Multiply the diagonals from top left to bottom right, and add these diagonals together.

$$\begin{pmatrix} a^* & b^\circ & c^\bullet \\ d & e^* & f^\circ \\ g & h & i^* \end{pmatrix} \begin{matrix} a & b \\ d^\bullet & e \\ g^\circ & h^\bullet \end{matrix} = a^*e^*i^* + b^\circ f^\circ g^\circ + c^\bullet d^\bullet h^\bullet$$

Then multiply the diagonals from bottom left to top right, and add these diagonals together.

$$\begin{pmatrix} a & b & c^* \\ d & e^* & f^\circ \\ g^* & h^\circ & i^\bullet \end{pmatrix} \begin{matrix} a^\circ & b^\bullet \\ d^\bullet & e \\ g & h \end{matrix} = g^*e^*c^* + h^\circ f^\circ a^\circ + i^\bullet d^\bullet b^\bullet$$

Subtract the first diagonal total minus the second diagonal total:

$$(= a^*e^*i^* + b^\circ f^\circ g^\circ + c^\bullet d^\bullet h^\bullet) - (= g^*e^*c^* + h^\circ f^\circ a^\circ + i^\bullet d^\bullet b^\bullet)$$

This gives the value of the determinant. To find the value of a variable, divide the value of the top determinant by the value of the bottom determinant.

1) $1x + 2y + 3z = 8.19$
2) $3y + 2z = 9.05$
3) $2x + 3y + 5z = 13.25$

$$x = \frac{\begin{pmatrix} 8.19 & 2 & 3 \\ 9.05 & 3 & 2 \\ 13.25 & 3 & 5 \end{pmatrix}}{\begin{pmatrix} 1 & 2 & 3 \\ 0 & 3 & 2 \\ 2 & 3 & 5 \end{pmatrix}}$$ solve each determinant using the method shown below

Multiply the diagonals from top left to bottom right, and add these diagonals together.

$$\begin{pmatrix} 8.19^* & 2^\circ & 3^\bullet \\ 9.05 & 3^* & 2^\circ \\ 13.25 & 3 & 5^* \end{pmatrix} \begin{matrix} 8.19 & 2 \\ 9.05^\bullet & 3 \\ 13.25^\circ & 3^\bullet \end{matrix}$$

$$= (8.19^*)(3^*)(5^*) + (2^\circ)(2^\circ)(13.25^\circ) + (3^\bullet)(9.05^\bullet)(3^\bullet)$$

Then multiply the diagonals from bottom left to top right, and add these diagonals together.

$$\begin{pmatrix} 8.19 & 2 & 3^* \\ 9.05 & 3^* & 2^\circ \\ 13.25^* & 3^\circ & 5^\bullet \end{pmatrix} \begin{matrix} 8.19^\circ & 2^\bullet \\ 9.05^\bullet & 3 \\ 13.25 & 3 \end{matrix}$$

$$= (13.25^*)(3^*)(3^*) + (3^\circ)(2^\circ)(8.19^\circ) + (5^\bullet)(9.05^\bullet)(2^\bullet)$$

Subtract the first diagonal total minus the second diagonal total:

$$(8.19^*)(3^*)(5^*) + (2^\circ)(2^\circ)(13.25^\circ) + (3^\bullet)(9.05^\bullet)(3^\bullet) = \ 257.30$$
$$- \ (13.25^*)(3^*)(3^*) + (3^\circ)(2^\circ)(8.19^\circ) + (5^\bullet)(9.05^\bullet)(2^\bullet) = \ ^-258.89$$

$$\overline{}$$

$$^-1.59$$

Do same multiplying and subtraction procedure for the bottom determinant to get $^-1$ as an answer. Now divide:

$$\frac{^-1.59}{^-1} = \$1.59/\text{lb of potato chips}$$

$$y = \frac{\begin{pmatrix} 1 & 8.19 & 3 \\ 0 & 9.05 & 2 \\ 2 & 13.25 & 5 \end{pmatrix}}{\begin{pmatrix} 1 & 2 & 3 \\ 0 & 3 & 2 \\ 2 & 3 & 5 \end{pmatrix}} = \frac{^-2.79}{^-1} = \$2.79/\text{lb of roast beef}$$

NOTE: The bottom determinant is always the same value for each letter.

$$z = \frac{\begin{pmatrix} 1 & 2 & 8.19 \\ 0 & 3 & 9.05 \\ 2 & 3 & 13.25 \end{pmatrix}}{\begin{pmatrix} 1 & 2 & 3 \\ 0 & 3 & 2 \\ 2 & 3 & 5 \end{pmatrix}} = \frac{^-.34}{^-1} = \$.34/\text{lb of apples}$$

Graphing

To graph an inequality, solve the inequality for y. This gets the inequality in **slope intercept form**, (for example $y < mx + b$). The point (0, b) is the y-intercept and m is the line's slope.

If the inequality solves to $x \geq$ **any number**, then the graph includes a **vertical line**.

If the inequality solves to $y \leq$ **any number**, then the graph includes a **horizontal line**.

When graphing a linear inequality, the line will be dotted if the inequality sign is $<$ or $>$. If the inequality signs are either $\geq$ or $\leq$, the line on the graph will be a solid line. Shade above the line when the inequality sign is $\geq$ or $>$. Shade below the line when the inequality sign is $<$ or $\leq$. For inequalities of the forms $x >$ number, $x \leq$ number, $x <$ number, or $x \geq$ number, draw a vertical line (solid or dotted). Shade to the right for $>$ or $\geq$. Shade to the left for $<$ or $\leq$.

Use these rules to graph and shade each inequality. The solution to a system of linear inequalities consists of the part of the graph that is shaded for each inequality. For instance, if the graph of one inequality is shaded with red, and the graph of another inequality is shaded with blue, then the overlapping area would be shaded purple. The purple area would be the points in the solution set of this system.

<u>Example</u>: Solve by graphing:

$$x + y \leq 6$$
$$x - 2y \leq 6$$

Solving the inequalities for y, they become:

$$y \leq -x + 6 \text{ [SA14]} \quad (y\text{-intercept of 6 and slope} = -1)$$
$$y \geq 1/2 x - 3 \quad (y \text{ intercept of } -3 \text{ and slope} = 1/2)$$

A graph with shading is shown below:

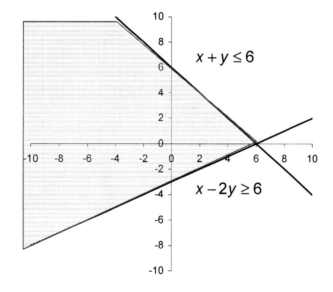

COMPETENCY 0008   **UNDERSTAND THE PROPERTIES OF QUADRATIC AND HIGHER-ORDER POLYNOMIAL RELATIONS AND FUNCTIONS.**

The standard form of a **quadratic function** is $f(x) = ax^2 + bx + c$.

Example: What patterns appear in a table for $y = x^2 - 5x + 6$?

x	y
0	6
1	2
2	0
3	0
4	2
5	6

We see that the values for y are **symmetrically** arranged.

Systems of quadratic equations can be solved by graphing but this method is sometimes inconclusive depending on the accuracy of the graphs and can also be cumbersome. It is recommended that either a substitution or elimination method be considered.

Sample problems:

Find the solution to the system of equations.

1. $$y^2 - x^2 = {}^-9$$
   $$2y = x - 3$$

     1. Use substitution method solving the second equation for x.

   $$2y = x - 3$$
   $$x = 2y + 3$$

     2. Substitute this into the first equation in place of (x).

   $$y^2 - (2y + 3)^2 = {}^-9$$

     3. Solve.

   $$y^2 - (4y^2 + 12y + 9) = {}^-9$$
   $$y^2 - 4y^2 - 12y - 9 = {}^-9$$
   $${}^-3y^2 - 12y - 9 = {}^-9$$
   $${}^-3y^2 - 12y = 0$$

     4. Factor.

   $${}^-3y(y + 4) = 0$$

$^-3y = 0$	$y + 4 = 0$	5. Set each factor equal to zero.
$y = 0$	$y = ^-4$	6. Use these values for y to solve for x.

$2y = x - 3$	$2y = x - 3$	7. Choose an equation.
$2(0) = x - 3$	$2(^-4) = x - 3$	8. Substitute.
$0 = x - 3$	$^-8 = x - 3$	
$x = 3$	$x = ^-5$	9. Write ordered pairs.

$(3,0)$ and $(^-5,^-4)$ satisfy the system of equations given.

2.
$$^-9x^2 + y^2 = 16$$
$$5x^2 + y^2 = 30$$

Use elimination to solve.

$$^-9x^2 + y^2 = 16$$
$$\underline{^-5x^2 - y^2 = ^-30}$$

1. Multiply second row by $^-1$.

2. Add.

$$^-14x^2 = ^-14$$
$$x^2 = 1$$
$$x = \pm 1$$

3. Divide by $^-14$.
4. Take the square root of both sides

$-9(1) + y^2 = 16$	$-9(-1) + y^2 = 16$
$^-9 + y^2 = 16$	$^-9 + y^2 = 16$
$y^2 = 25$	$y^2 = 25$
$y = \pm 5$	$y = \pm 5$
$(1, \pm 5)$	$(^-1, \pm 5)$

5. Substitute both values of x into the equation.

6. Take the square root of both sides.

7. Write the ordered pairs.

$(1, \pm 5)$ and $(-1, \pm 5)$

Satisfy the system of equations

- To factor the **sum or the difference of perfect cubes**, follow this procedure:

a. Factor out any greatest common factor (GCF).

b. Make a parentheses for a binomial (2 terms) followed by a trinomial (3 terms).

c. The sign in the first parentheses is the same as the sign in the problem. The difference of cubes will have a "-" sign in the first parentheses. The sum of cubes will use a "+".

d. The first sign in the second parentheses is the opposite of the sign in the first parentheses. The second sign in the other parentheses is always a "+".

e. Determine what would be cubed to equal each term of the problem. Put those expressions in the first parentheses.

f. To make the 3 terms of the trinomial, think square - product - square. Looking at the binomial, square the first term. This is the trinomial's first term. Looking at the binomial, find the product of the two terms, ignoring the signs. This is the trinomial's second term. Looking at the binomial, square the third term. This is the trinomial's third term. Except in rare instances, the trinomial does not factor again.

Factor completely:

1.

$16x^3 + 54y^3$

$2(8x^3 + 27y^3)$ ← GCF

$2(\quad + \quad)(\quad - \quad + \quad)$ ← signs

$2(2x + 3y)(\quad - \quad + \quad)$ ← what is cubed to equal $8x^3$ or $27y^3$

$2(2x + 3y)(4x^2 - 6xy + 9y^2)$ ← square-product-square

2.

$64a^3 - 125b^3$

$(\quad - \quad)(\quad + \quad + \quad) \leftarrow$ signs

$(4a - 5b)(\quad + \quad + \quad) \leftarrow$ what is cubed to equal $64a^3$ or $125b^3$

$(4a - 5b)(16a^2 + 20ab + 25b^2) \leftarrow$ square-product-square

3.

$27x^{27} + 343y^{12} = (3x^9 + 7y^4)(9x^{18} - 21x^9y4 + 49y^8)$

Note: The coefficient 27 is different from the exponent 27.

Try These:

1. $216x^3 - 125y^3$
2. $4a^3 - 32b^3$
3. $40x^{29} + 135x^2y^3$

To factor a polynomial, follow these steps:

a. **Factor out any GCF** (greatest common factor)

b. For a binomial (2 terms), check to see if the problem is the **difference of perfect squares**. If both factors are perfect squares, then it factors this way:

$$a^2 - b^2 = (a - b)(a + b)$$

If the problem is not the difference of perfect squares, then check to see if the problem is either the sum or difference of perfect cubes.

$x^3 - 8y^3 = (x - 2y)(x^2 + 2xy + 4y^2) \qquad \leftarrow$ difference

[SA15]

$64a^3 + 27b^3 = (4a + 3b)(16a^2 - 12ab + 9b^2) \leftarrow$ sum

[SA16]c. Trinomials can be perfect squares. Trinomials can be factored into 2 binomials (un-FOILing). Be sure the terms of the trinomial are in descending order. If the last sign of the trinomial is a "+," then the signs in the parentheses will be the same as the sign in front of the second term of the trinomial. If the last sign of the trinomial is a "−," then there will be one "+" and one "−" in the two parentheses. The first term of the trinomial can be factored to equal the first terms of the two factors. The last term of the trinomial can be factored to equal the last terms of the two factors. Work backwards to determine the correct factors to multiply together to get the correct center term.

Factor completely:

1. $4x^2 - 25y^2$ [SA17]
2. $6b^2 - 2b - 8$

Answers:

1. No GCF; this is the difference of perfect squares.

$$4x^2 - 25y^2 = (2x - 5y)(2x + 5y)$$

2. GCF of 2; Try to factor into 2 binomials:

$$6b^2 - 2b - 8 = 2(3b^2 - b - 4)$$

Since the last sign of the trinomial is a "−," the signs in the factors are one "+," and one "−." $3b^2$ factors into $3b$ and b. Find factors of 4: 1 & 4; 2 & 2.

$$6b^2 - 2b - 8 = 2(3b^2 - b - 4) = 2(3b - 4)(b + 1)$$

Solve Quadratics

A **quadratic equation** is written in the form $ax^2 + bx + c = 0$. To solve a quadratic equation by factoring, at least one of the factors must equal zero.

Example:

Solve the equation.

$x^2 + 10x - 24 = 0$

$(x + 12)(x - 2) = 0$ Factor.   [SA18]

$x + 12 = 0$ or $x - 2 = 0$ Set each factor equal to 0.

$x = -12$ $x = 2$ Solve.

Check:

$$x^2 + 10x - 24 = 0$$

$$(-12)^2 + 10(-12) - 24 = 0 \qquad (2)^2 + 10(2) - 24 = 0 \text{ [SA19]}$$

$$144 - 120 - 24 = 0 \qquad 4 + 20 - 24 = 0$$

$$0 = 0 \qquad 0 = 0$$

A quadratic equation that cannot be solved by factoring can be solved by **completing the square**.

<u>Example</u>:

Solve the equation.

$$x^2 - 6x + 8 = 0$$

$$x^2 - 6x = {}^-8 \qquad \text{Move the constant to the right side.}$$

$$x^2 - 6x + 9 = {}^-8 + 9 \qquad \text{Add the square of half the coefficient of } x \text{ to both sides.}$$

$$(x - 3)^2 = 1 \qquad \text{Write the left side as a perfect square.}$$

$$x - 3 = \pm\sqrt{1} \qquad \text{Take the square root of both sides.}$$

$$x - 3 = 1 \qquad x - 3 = {}^-1 \qquad \text{Solve.}$$

$$x = 4 \qquad x = 2$$

[SA20]

Check:

$$x^2 - 6x + 8 = 0$$

$$4^2 - 6(4) + 8 = 0 \qquad 2^2 - 6(2) + 8 = 0$$

$$16 - 24 + 8 = 0 \qquad 4 - 12 + 8 = 0$$

$$0 = 0 \qquad 0 = 0$$

The general technique for **graphing quadratics** is the same as for graphing linear equations. Graphing quadratic equations, however, results in a parabola instead of a straight line.

Example:

Graph $y = 3x^2 + x - 2$.

x	$y = 3x^2 + x - 2$
$^-2$	8
$^-1$	0
0	$^-2$
1	2
2	12

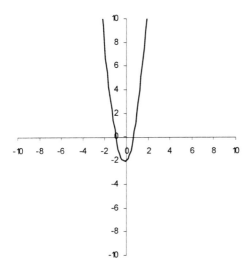

To solve a quadratic equation using the quadratic formula, be sure that your equation is in the form $ax^2 + bx + c = 0$. Substitute these values into the formula:

$$x = \frac{-b \pm \sqrt{b^2 - 4ac}}{2a}$$

Example:

Solve the equation.

$$3x^2 = 7 + 2x \rightarrow 3x^2 - 2x - 7 = 0$$

$$a = 3 \quad b = {}^-2 \quad c = {}^-7$$

$$x = \frac{-({}^-2) \pm \sqrt{({}^-2)^2 - 4(3)({}^-7)}}{2(3)}$$

$$x = \frac{2 \pm \sqrt{4 + 84}}{6}$$

$$x = \frac{2 \pm \sqrt{88}}{6}$$

$$x = \frac{2 \pm 2\sqrt{22}}{6}$$

$$x = \frac{1 \pm \sqrt{22}}{3}$$

Word Problems – Quadratics

Some word problems will give a quadratic equation to be solved. When the quadratic equation is found, set it equal to zero and solve the equation by factoring or the quadratic formula. Examples of this type of problem follow.

Example:
Ashland (A) is a certain distance north of Belmont (B). The distance from Belmont east to Carlisle (C) is 5 miles more than the distance from Ashland to Belmont. The distance from Ashland to Carlisle is 10 miles more than the distance from Alberta to Belmont. How far is Ashland from Carlisle?

Solution:
Since north and east form a right angle, these distances are the lengths of the legs of a right triangle. If the distance from Ashland to Belmont is x, then from Belmont to Carlisle is $x + 5$, and the distance from Ashland to Carlisle is $x + 10$.

The equation is: $\qquad AB^2 + BC^2 = AC^2$

$$x^2 + (x+5)^2 = (x+10)^2$$

$$x^2 + x^2 + 10x + 25 = x^2 + 20x + 100$$

$$2x^2 + 10x + 25 = x^2 + 20x + 100$$

$$x^2 - 10x - 75 = 0$$

$(x - 15)(x + 5) = 0$ Distance cannot be negative.

$x = 15$ Distance from Ashland to Belmont.

$x + 5 = 20$ Distance from Belmont to Carlisle.

$x + 10 = 25$ Distance from Ashland to Carlisle.

Example:
The square of a number is equal to 6 more than the original number. Find the original number.

Solution: If x = original number, then the equation is:

$x^2 = 6 + x$ Set this equal to zero.

$x^2 - x - 6 = 0$ Now factor.

$(x - 3)(x + 2) = 0$

$x = 3$ or $x = {}^- 2$ There are 2 solutions, 3 or $^-$2.

Try these:

1. One side of a right triangle is 1 less than twice the shortest side, while the third side of the triangle is 1 more than twice the shortest side. Find all 3 sides.

2. Twice the square of a number equals 2 less than 5 times the number. Find the number(s).

Some word problems can be solved by setting up a quadratic equation or inequality. Examples of this type could be problems that deal with finding a maximum area.

Example:

A family wants to enclose 3 sides of a rectangular garden with 200 feet of fence. In order to have a garden with an area of **at least** 4800 square feet, find the dimensions of the garden. Assume that a wall or a fence already borders the fourth side of the garden

Solution:
Let x = distance from the wall

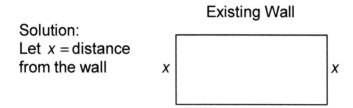

Existing Wall

Then 2x feet of fence is used for these 2 sides. The remaining side of the garden would use the rest of the 200 feet of fence, that is, $200 - 2x$ feet of fence. Therefore the width of the garden is x feet and the length is $200 - 2x$ ft.

The area, $200x - 2x^2$, needs to be greater than or equal to 4800 sq. ft. So, this problem uses the inequality $4800 \le 200x - 2x^2$. This becomes $2x^2 - 200x + 4800 \le 0$. Solving this, we get:

$$200x - 2x^2 \ge 4800$$
$$-2x^2 + 200x - 4800 \ge 0$$
$$2\left(-x^2 + 100x - 2400\right) \ge 0$$
$$-x^2 + 100x - 2400 \ge 0$$
$$(-x + 60)(x - 40) \ge 0$$
$$-x + 60 \ge 0$$
$$-x \ge -60$$
$$x \le 60$$
$$x - 40 \ge 0$$
$$x \ge 40$$

So the area will be at least 4800 square feet if the width of the garden is from 40 up to 60 feet. (The length of the rectangle would vary from 120 feet to 80 feet depending on the width of the garden.)

Graphing

Quadratic equations can be used to model different real life situations. The graphs of these quadratics can be used to determine information about this real life situation.

<u>Example</u>:
The height of a projectile fired upward at a velocity of v meters per second from an original height of h meters is $y = h + vx - 4.9x^2$.
If a rocket is fired from an original height of 250 meters with an original velocity of 4800 meters per second, find the approximate time the rocket would drop to sea level (a height of 0).

Solution:

The equation for this problem is: $y = 250 + 4800x - 4.9x^2$. If the height at sea level is zero, then $y = 0$ so $0 = 250 + 4800x - 4.9x^2$. Solving this for x could be done by using the quadratic formula. In addition, the approximate time in x seconds until the rocket would be at sea level could be estimated by looking at the graph. When the y value of the graph goes from positive to negative then there is a root (also called the solution or x intercept) in that interval.

$$x = \frac{-4800 \pm \sqrt{4800^2 - 4(-4.9)(250)}}{2(-4.9)} \approx 980 \text{ or } -0.05 \text{ seconds}$$

Since the time has to be positive, it will be about 980 seconds until the rocket is at sea level.

To graph an inequality, graph the quadratic as if it was an equation; however, if the inequality has just a > or < sign, then make the curve itself dotted. Shade above the curve for > or ≥. Shade below the curve for < or ≤.

Examples:

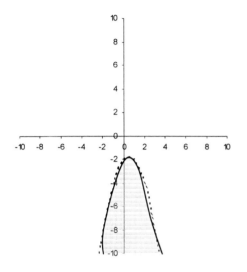

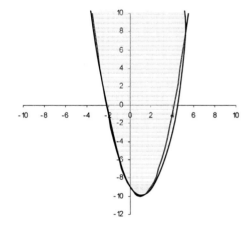

Solving Quadratics

To solve **a quadratic equation** (with x^2), rewrite the equation into the form:

$$ax^2 + b\,x + c = 0 \text{ or } y = ax^2 + b\,x + c$$

where a, b, and c are real numbers. Then substitute the values of a, b, and c into the quadratic formula:

$$x = \frac{-b \pm \sqrt{b^2 - 4ac}}{2a}$$

Simplify the result to find the answers. (Remember, there could be 2 real answers, one real answer, or 2 complex answers that include "i").

To solve a quadratic inequality (with x^2), solve for y. The axis of symmetry is located at $x = {}^-b/2a$. Find coordinates of points to each side of the axis of symmetry. Graph the parabola as a dotted line if the inequality sign is either $<$ or $>$. Graph the parabola as a solid line if the inequality sign is either $\leq$ or $\geq$. Shade above the parabola if the sign is $\geq$ or $>$. Shade below the parabola if the sign is $\leq$ or $<$.

<u>Example</u>: Solve: $8x^2 - 10x - 3 = 0$

In this equation $a = 8$, $b = {}^-10$, and $c = {}^-3$.

Substituting these into the quadratic equation, it becomes:

$$x = \frac{{}^-({}^-10) \pm \sqrt{({}^-10)^2 - 4(8)({}^-3)}}{2(8)} = \frac{10 \pm \sqrt{100 + 96}}{16}$$

$$x = \frac{10 \pm \sqrt{196}}{16} = \frac{10 \pm 14}{16} = 24/16 = 3/2 \text{ or } {}^-4/16 = {}^-1/4$$

<u>Check:</u>
$$x = -\frac{1}{4}$$

$$\frac{1}{2} + \frac{10}{4} - 3 = 0 \qquad \text{Both Check}$$

$$3 - 3 = 0$$

<u>Example</u>: Solve and graph : $y > x^2 + 4x - 5$.

The axis of symmetry is located at $x = {}^-b/2a$. Substituting 4 for b, and 1 for a, this formula becomes:

$$x = {}^-(4)/2(1) = {}^-4/2 = {}^-2$$

Find coordinates of points to each side of $x = {}^-2$.

x	y
$^-5$	0
$^-4$	$^-5$
$^-3$	$^-8$
$^-2$	$^-9$
$^-1$	$^-8$
0	$^-5$
1	0

Graph these points to form a parabola. Draw it as a dotted line. Since a greater than sign is used, shade above and inside the parabola.

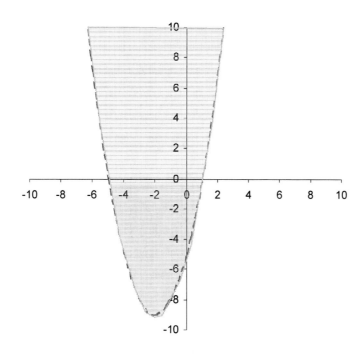

The discriminant of a quadratic equation is the part of the quadratic formula that is usually inside the radical sign, $b^2 - 4ac$.

$$x = \frac{-b \pm \sqrt{b^2 - 4ac}}{2a}$$

The radical sign is NOT part of the discriminant!! Determine the value of the discriminant by substituting the values of a, b, and c from $ax^2 + bx + c = 0$.

-If the value of the discriminant is **any negative number**, then there are **two complex roots** including "i."

-If the value of the discriminant is **zero**, then there is only **1 real rational root**. This would be a double root.

-If the value of the discriminant is **any positive number that is also a perfect square**, then there are **two real rational roots.** (There are no longer any radical signs.)

-If the value of the discriminant is **any positive number that is NOT a perfect square**, then there are **two real irrational roots.** (There are still unsimplified radical signs.)

Example:

Find the value of the discriminant for the following equations. Then determine the number and nature of the solutions of that quadratic equation.

$$2x^2 - 5x + 6 = 0$$

$a = 2$, $b = {}^-5$, $c = 6$ so $b^2 - 4ac = ({}^-5)^2 - 4(2)(6) = 25 - 48 = {}^-23$.

Since ${}^-23$ is a negative number, there are **two complex roots** including "i".

$$x = \frac{5}{4} + \frac{i\sqrt{23}}{4}, \quad x = \frac{5}{4} - \frac{i\sqrt{23}}{4}$$

$$3x^2 - 12x + 12 = 0$$

$a = 3$, $b = {}^-12$, $c = 12$ so $b^2 - 4ac = ({}^-12)^2 - 4(3)(12) = 144 - 144 = 0$

Since 0 is the value of the discriminant, there is only
1 real rational root.

$$x=2.$$

$$6x^2 - x - 2 = 0$$

$a = 6,\ b = {}^-1,\ c = {}^-2$ so $b^2 - 4ac = ({}^-1)^2 - 4(6)({}^-2) = 1 + 48 = 49$.

Since 49 is positive and is also a perfect square $(\sqrt{49}) = 7$, then there are **two real rational roots.**

$$x = \frac{2}{3},\quad x = -\frac{1}{2}$$

Try these. (The answers are in the Answer Key to Practice Problems):

1. $6x^2 - 7x - 8 = 0$
2. $10x^2 - x - 2 = 0$
3. $25x^2 - 80x + 64 = 0$

Determine a Quadratic from Known Roots

Follow these steps to write a quadratic equation from its roots:

1. Add the roots together. The answer is their **sum**. Multiply the roots together. The answer is their **product**.
2. A quadratic equation can be written using the sum and product like this:

$$x^2 + \textbf{(opposite of the sum)}x + \textbf{product} = 0$$

3. If there are any fractions in the equation, multiply every term by the common denominator to eliminate the fractions. This is the quadratic equation.
4. If a quadratic equation has only 1 root, use it twice and follow the first 3 steps above.

Example:
Find a quadratic equation with roots of 4 and ⁻9.

Solutions:
The sum of 4 and ⁻9 is ⁻5. The product of 4 and ⁻9 is ⁻36.
The equation would be:

$$x^2 + \text{(opposite of the sum)}x + \text{product} = 0$$
$$x^2 + 5x - 36 = 0$$

Find a quadratic equation with roots of $5 + 2i$ and $5 - 2i$.

Solutions:
The sum of $5 + 2i$ and $5 - 2i$ is 10. The product of $5 + 2i$ and $5 - 2i$
is $25 - 4i^2 = 25 + 4 = 29$.
The equation would be:

$$x^2 + \text{(opposite of the sum)}x + \text{product} = 0$$
$$x^2 - 10x + 29 = 0$$

Find a quadratic equation with roots of $2/3$ and $^-3/4$.

Solutions:
The sum of $2/3$ and $^-3/4$ is $^-1/12$. The product of
$2/3$ and $^-3/4$ is $^-1/2$.
The equation would be :

$$x^2 + \text{(opposite of the sum)}x + \text{product} = 0$$
$$x^2 + 1/12\, x - 1/2 = 0$$

Common denominator = 12, so multiply by 12.

$$12(x^2 + 1/12\, x - 1/2 = 0$$
$$12x^2 + 1x - 6 = 0$$
$$12x^2 + x - 6 = 0$$

Try these:

1. Find a quadratic equation with a root of 5.
2. Find a quadratic equation with roots of $8/5$ and $^-6/5$.
3. Find a quadratic equation with roots of 12 and $^-3$.

Finding Zeros of a Function

Synthetic division can be used to find the value of a function at any value of x. To do this, divide the value of x into the coefficients of the function.

(Remember that coefficients of missing terms, like x^2 below, must be included.) The remainder of the synthetic division is the value of the function. If $f(x) = x^3 - 6x + 4$, then to find the value of the function at $x = 3$, use synthetic division:

Note the 0 for the missing x^2 term.

$$
\begin{array}{r|rrrr}
3 & 1 & 0 & ^-6 & 4 \\
  &   & 3 & 9 & 9 \\
\hline
  & 1 & 3 & 3 & 13
\end{array}
$$
← This is the value of the function.

Therefore, (3, 13) is a point on the graph.

Example: Find values of the function at $x = {}^-5$ if
$$f(x) = 2x^5 - 4x^3 + 3x^2 - 9x + 10.$$

Note the 0 below for the missing x^4 term.

Synthetic division:

$$
\begin{array}{r|rrrrrr}
^-5 & 2 & 0 & ^-4 & 3 & ^-9 & 10 \\
    &   & ^-10 & 50 & ^-230 & 1135 & ^-5630 \\
\hline
    & 2 & ^-10 & 46 & ^-227 & 1126 & ^-5620
\end{array}
$$
← This is the value of the function if $x = {}^-5$.

Therefore, ($^-5$, $^-5620$) is a point on the graph.

Note that if $x = {}^-5$, the same value of the function can also be found by substituting $^-5$ in place of x in the function.

$$f(^-5) = 2(^-5)^5 - 4(^-5)^3 + 3(^-5)^2 - 9(^-5) + 10$$
$$= 2(^-3125) - 4(^-125) + 3(25) - 9(^-5) + 10$$
$$= {}^-6250 + 500 + 75 + 45 + 10 = {}^-5620$$

Therefore, ($^-5$, $^-5620$) is still a point on the graph.

To determine if $(x-a)$ or $(x+a)$ is a factor of a polynomial, do a synthetic division, dividing by the opposite of the number inside the parentheses. To see if $(x-5)$ is a factor of a polynomial, divide it by 5. If the remainder of the synthetic division is zero, then the binomial is a factor of the polynomial.

If $f(x) = x^3 - 6x + 4$, determine if $(x-1)$ is a factor of $f(x)$. Use synthetic division and divide by 1:

Note the 0 for the missing x^2 term.

$$
1 \overline{)\begin{array}{cccc} 1 & 0 & {}^-6 & 4 \\ & 1 & 1 & {}^-5 \end{array}}
$$

1   1   ‾5   ‾1 ← This is the remainder of the function.

Therefore, $(x-1)$ is **not** a factor of f(x).

If $f(x) = x^3 - 6x + 4$, determine if $(x-2)$ is a factor of f(x). Use synthetic division and divide by 2:

$$
2 \overline{)\begin{array}{cccc} 1 & 0 & {}^-6 & 4 \\ & 2 & 4 & {}^-4 \end{array}}
$$

1   2   ‾2   0 ← This is the remainder of the function.

Therefore, $(x-2)$ **is** a factor of f(x).

The converse of this is also true. If you divide by k in any synthetic division and get a remainder of zero for the division, then $(x-k)$ is a factor of the polynomial. Similarly, if you divide by ^-k in any synthetic division and get a remainder of zero for the division, then $(x+k)$ is a factor of the polynomial.

Divide $2x^3 - 6x - 104$ by 4. What is your conclusion?

$$
4 \overline{)\begin{array}{cccc} 2 & 0 & {}^-6 & {}^-104 \\ & 8 & 32 & 104 \end{array}}
$$

2   8   26   0 ← This is the remainder of the function.

Since the remainder is **0**, then $(x-4)$ is a factor.

Given any polynomial, be sure that the exponents of the terms are in descending order. List out all the factors of the first term's coefficient, and of the constant in the last term. Make a list of fractions by putting each of the factors of the last term's coefficient over each of the factors of the first term. Reduce fractions when possible. Put a $\pm$ in front of each fraction. This list of fractions is a list of the only possible rational roots of a function. If the polynomial is of degree **n**, then at most n of these will actually be roots of the polynomial.

Example: List the possible rational roots for the function $f(x) = x^2 - 5x + 4$.

$$\pm \frac{\text{factors of 4}}{\text{factors of 1}} = \pm 1, 2, 4 \leftarrow 6 \text{ possible rational roots}$$

Example: List the possible rational roots for the function $f(x) = 6x^2 - 5x - 4$.

Make fractions of the form :

possible
rational roots $= \pm \dfrac{\text{factors of 4}}{\text{factors of 6}} = \pm \dfrac{1,2,4}{1,2,3,6} =$

$\pm \dfrac{1}{2}, \dfrac{1}{3}, \dfrac{1}{6}, \dfrac{2}{3}, \dfrac{4}{3}, 1, 2, 4$ are the only 16 rational
numbers that could be roots.

Since this equation is of degree 2, there are, at most, 2 rational roots. (They happen to be 4/3 and $^-1/2$.)

Descarte's Rule of signs can help to determine how many positive real roots or how many negative real roots a function would have. Given any polynomial, be sure that the exponents of the terms are in descending order. Count the number of successive terms of the polynomial where there is a sign change. The number of positive roots will be equal to the number of sign changes or will be less than the number of sign changes by a multiple of 2. For example,

$$y = 2x^5 + 3x^4 - 6x^3 + 4x^2 + 8x - 9 \quad \text{has 3 sign}$$
$$\rightarrow 1 \quad \rightarrow 2 \qquad \rightarrow 3 \quad \text{changes.}$$

That means that this equation will have either 3 positive roots or 1 positive root.

The equation:

$$y = 4x^6 - 5x^5 + 6x^4 - 3x^3 + 2x^2 + 8x - 10 \quad \text{has 5 sign}$$
$$\rightarrow 1 \quad \rightarrow 2 \quad \rightarrow 3 \quad \rightarrow 4 \qquad \rightarrow 5 \quad \text{changes.}$$

This equation will have either 5 positive roots (equal to the number of sign changes) or 3 positive roots (2 less) or only 1 positive root (4 less).

The equation:

$$y = x^8 - 3x^5 - 6x^4 - 3x^3 + 2x^2 - 8x + 10 \quad \text{has 4 sign}$$
$$\rightarrow 1 \qquad\qquad\qquad \rightarrow 2 \quad \rightarrow 3 \quad \rightarrow 4 \quad \text{changes.}$$

This equation will have either 4 positive roots (equal to the number of sign changes) or 2 positive roots (2 less) or no positive roots(4 less).

The second part of Descarte's Rule of signs also requires that terms are in descending order of exponents. Next, look at the equation and change the signs of the terms that have an odd exponent. Then count the number of sign changes in successive terms of the new polynomial. The number of negative terms will be equal to the number of sign changes, or will be less than the number of sign changes by a multiple of 2.

For example, given the equation:

$$y = 2x^5 + 3x^4 - 6x^3 + 4x^2 + 8x - 9$$

Change the signs of the terms with odd exponents.

$$y = -2x^5 + 3x^4 + 6x^3 + 4x^2 - 8x - 9$$

Now count the number of sign changes in this equation.

$$y = -2x^5 + 3x^4 + 6x^3 + 4x^2 - 8x - 9 \quad \text{has 2 sign}$$
$$\rightarrow 1 \qquad\qquad\qquad \rightarrow 2 \qquad \text{changes.}$$

This tells you that there are 2 negative roots or 0 negative roots (2 less).

Example: Determine the number of positive or negative real roots for the equation:

$$y = x^3 + 9x^2 + 23x + 15$$

This equation is of degree 3, so it has, at most, 3 roots. Look at the equation. There are 0 sign changes. This means there are 0 positive roots. To check for negative roots, change the signs of the terms with odd exponents. The equation becomes:

$$y = -x^3 + 9x^2 - 23x + 15 \quad \text{Now count sign changes.}$$
$$\to 1 \quad \to 2 \quad \to 3 \quad \text{There are 3 sign changes.}$$

This means there are either 3 negative roots or only 1 negative root.

To find points on the graph of a polynomial, substitute desired values in place of x and solve for the corresponding y value of that point on the graph. A second way to do the same thing is to do a synthetic division, dividing by the x value of the desired point. The remainder at the end of the synthetic division is the y value of the point. Find a group of points to plot, then graph and connect the points from left to right on the graph. The y intercept will always have a y value equal to the constant of the equation.

PARABOLAS-A parabola is a set of all points in a plane that are equidistant from a fixed point (focus) and a line (directrix).

FORM OF EQUATION $\quad y = a(x - h)^2 + k \qquad x = a(y - k)^2 + h$

IDENTIFICATION $\quad x^2$ term, y not squared $\quad y^2$ term, x not squared

SKETCH OF GRAPH

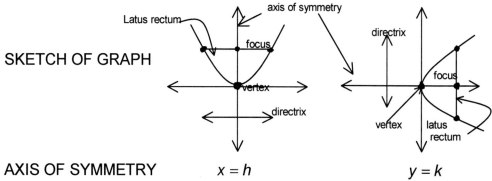

AXIS OF SYMMETRY $\qquad x = h \qquad\qquad\qquad y = k$

-A line through the vertex and focus upon which the parabola is symmetric.

VERTEX	(h,k)	(h,k)

-The point where the parabola intersects the axis of symmetry.

FOCUS	$(h, k + 1/4a)$	$(h + 1/4a, k)$

DIRECTRIX	$y = k - 1/4a$	$x = h - 1/4a$

DIRECTION OF OPENING	up if $a > 0$, down if $a < 0$	right if $a > 0$, left if $a < 0$

LENGTH OF LATUS RECTUM	$\left| 1/a \right|$	$\left| 1/a \right|$

-A chord through the focus, perpendicular to the axis of symmetry, with endpoints on the parabola.

Sample Problem:

1. Find all identifying features of $y = {}^{-}3x^2 + 6x - 1$.

First, the equation must be put into the general form

$y = a(x - h)^2 + k$.

$y = {}^{-}3x^2 + 6x - 1$ 1. Begin by completing the square.

$\quad = {}^{-}3(x^2 - 2x + 1) - 1 + 3$

$\quad = {}^{-}3(x - 1)^2 + 2$ 2. Using the general form of the equation to identify known variables.

$a = {}^{-}3 \quad h = 1 \quad k = 2$

axis of symmetry: $x = 1$
vertex: $(1,2)$
focus: $(1, 1\frac{1}{4})$

directrix: $y = 2\frac{3}{4}$
direction of opening: down since $a < 0$
length of latus rectum: $1/3$

ELLIPSE

FORM OF EQUATION $\dfrac{(x-h)^2}{a^2}+\dfrac{(y-k)^2}{b^2}=1$ $\dfrac{(x-h)^2}{b^2}+\dfrac{(y-k)^2}{a^2}=1$

(for ellipses where $a^2 > b^2$).

where $b^2 = a^2 - c^2$ where $b^2 = a^2 - c^2$

IDENTIFICATION   horizontal major axis   vertical major axis

SKETCH

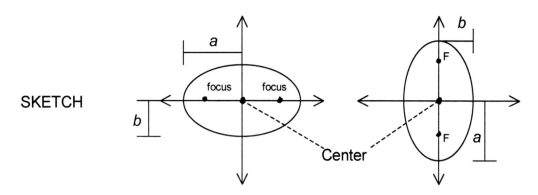

	horizontal major axis	vertical major axis
CENTER	(h,k)	(h,k)
FOCI	$(h\pm c,k)$	$(h,k\pm c)$
MAJOR AXIS LENGTH	$2a$	$2a$
MINOR AXIS LENGTH	$2b$	$2b$

Sample Problem:

Find all identifying features of the ellipse $2x^2 + y^2 - 4x + 8y - 6 = 0$.

First, begin by writing the equation in standard form for an ellipse.

$2x^2 + y^2 - 4x + 8y - 6 = 0$ 1. Complete the square for each variable.

$2(x^2 - 2x + 1) + (y^2 + 8y + 16) = 6 + 2(1) + 16$

$2(x-1)^2 + (y+4)^2 = 24$ 2. Divide both sides by 24.

$\dfrac{(x-1)^2}{12} + \dfrac{(y+4)^2}{24} = 1$ 3. Now the equation is in standard form.

Identify known variables: $h = 1 \quad k = {}^-4 \qquad a = \sqrt{24}$ or $2\sqrt{6}$

$\qquad\qquad\qquad\qquad\qquad\quad b = \sqrt{12}$ or $2\sqrt{3} \quad c = 2\sqrt{3}$

Identification: vertical major axis

Center: $(1, {}^-4)$

Foci: $(1, {}^-4 \pm 2\sqrt{3})$

Major axis: $4\sqrt{6}$

Minor axis: $4\sqrt{3}$

HYPERBOLA

FORM OF EQUATION	$\dfrac{(x-h)^2}{a^2} - \dfrac{(y-k)^2}{b^2} = 1$ where $c^2 = a^2 + b^2$	$\dfrac{(y-k)^2}{a^2} - \dfrac{(x-h)^2}{b^2} = 1$ where $c^2 = a^2 + b^2$
IDENTIFICATION	horizontal transverse axis (y^2 is negative)	vertical transverse axis (x^2 is negative)
SKETCH	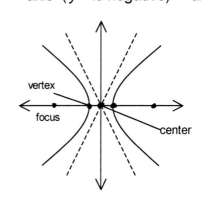	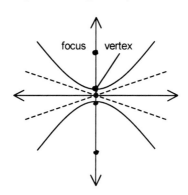
SLOPE OF ASYMPTOTES	$\pm(b/a)$	$\pm(a/b)$
TRANSVERSE AXIS (endpoints are vertices of the hyperbola and go through the center)	$2a$ -on y axis	$2a$ -on x axis
CONJUGATE AXIS (perpendicular to transverse axis at center)	$2b$, -on y axis	$2b$, -on x axis
CENTER	(h,k)	(h,k)
FOCI	$(h \pm c, k)$	$(h, k \pm c)$
VERTICES	$(h \pm a, k)$	$(h, k \pm a)$

Sample Problem:

Find all the identifying features of a hyperbola given its equation.

$$\frac{(x+3)^2}{4} - \frac{(y-4)^2}{16} = 1$$

Identify all known variables: $h = {}^-3 \quad k = 4 \quad a = 2 \quad b = 4 \quad c = 2\sqrt{5}$

Slope of asymptotes: $\pm 4/2$ or ± 2
Transverse axis: 4 units long
Conjugate axis: 8 units long
Center: $({}^-3,4)$
Foci: $({}^-3 \pm 2\sqrt{5},4)$
Vertices: $({}^-1,4)$ and $({}^-5,4)$

Circles

The equation of a circle with its center at (h,k) and a radius r units is:

$$(x-h)^2 + (y-k)^2 = r^2$$

Sample Problem:

1. Given the equation $x^2 + y^2 = 9$, find the center and the radius of the circle. Then graph the equation.

First, writing the equation in standard circle form gives:

$$(x-0)^2 + (y-0)^2 = 3^2$$

therefore, the center is (0,0) and the radius is 3 units.

Sketch the circle:

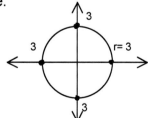

2. Given the equation $x^2 + y^2 - 3x + 8y - 20 = 0$, find the center and the radius. Then graph the circle.

First, write the equation in standard circle form by completing the square for both variables.

$x^2 + y^2 - 3x + 8y - 20 = 0$ 1. Complete the squares.

$(x^2 - 3x + 9/4) + (y^2 + 8y + 16) = 20 + 9/4 + 16$

$(x - 3/2)^2 + (y + 4)^2 = 153/4$

The center is $(3/2, {}^-4)$ and the radius is $\dfrac{\sqrt{153}}{2}$ or $\dfrac{3\sqrt{17}}{2}$.

Graph the circle.

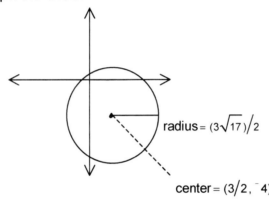

To write the equation given the center and the radius use the standard form of the equation of a circle:

$$(x - h)^2 + (y - k)^2 = r^2$$

Sample problems:

Given the center and radius, write the equation of the circle.

1. Center ($^-1,4$); radius 11

    $(x - h)^2 + (y - k)^2 = r^2$ 1. Write standard equation.

    $(x - (^-1))^2 + (y - (4))^2 = 11^2$ 2. Substitute.

    $(x + 1)^2 + (y - 4)^2 = 121$ 3. Simplify.

2. Center ($\sqrt{3}, ^-1/2$); radius $= 5\sqrt{2}$

    $(x - h)^2 + (y - k)^2 = r^2$ 1. Write standard equation.

    $(x - \sqrt{3})^2 + (y - (^-1/2))^2 = (5\sqrt{2})^2$ 2. Substitute.

    $(x - \sqrt{3})^2 + (y + 1/2)^2 = 50$ 3. Simplify.

COMPETENCY 0009   UNDERSTAND THE PRINCIPLES AND PROPERTIES OF RATIONAL, RADICAL, PIECEWISE, AND ABSOLUTE VALUE FUNCTIONS.

In order **to add or subtract** rational expressions, they must have a common denominator. If they don't have a common denominator, then factor the denominators to determine what factors are missing from each denominator to make the LCD. Multiply both numerator and denominator by the missing factor(s). Once the fractions have a common denominator, add or subtract their numerators, but keep the common denominator the same. Factor the numerator if possible and reduce if there are any factors that can be cancelled.

In order **to multiply** rational expressions, they do not have to have a common denominator. If you factor each numerator and denominator, you can cancel out any factor that occurs in both the numerator and denominator. Then multiply the remaining factors of the numerator together. Last multiply the remaining factors of the denominator together.

In order **to divide** rational expressions, the problem must be re-written as the first fraction multiplied times the inverse of the second fraction. Once the problem has been written as a multiplication, factor each numerator and denominator. Cancel out any factor that occurs in both the numerator and denominator. Then multiply the remaining factors of the numerator together. Last multiply the remaining factors of the denominator together.

1. $\dfrac{5}{x^2-9} - \dfrac{2}{x^2+4x+3} = \dfrac{5}{(x-3)(x+3)} - \dfrac{2}{(x+3)(x+1)} =$

$$\dfrac{5(x+1)}{(x+1)(x-3)(x+3)} - \dfrac{2(x-3)}{(x+3)(x+1)(x-3)} = \dfrac{3x+11}{(x-3)(x+3)(x+1)}$$

2. $\dfrac{x^2-2x-24}{x^2+6x+8} \times \dfrac{x^2+3x+2}{x^2-13x+42} = \dfrac{(x-6)(x+4)}{(x+4)(x+2)} \times \dfrac{(x+2)(x+1)}{(x-7)(x-6)} = \dfrac{x+1}{x-7}$

Practice Problems:

1. $\dfrac{6}{x^2-1} + \dfrac{8}{x^2+7x+6}$

2. $\dfrac{x^2-9}{x^2-4} \div \dfrac{x^2+8x+15}{x^3+8}$

Rational Expressions – Addition and Subtraction

- In order to **add or subtract rational expressions**, they must have a common denominator. If they don't have a common denominator, then factor the denominators to determine what factors are missing from each denominator to make the LCD. Multiply both numerator and denominator by the missing factor(s). Once the fractions have a common denominator, add or subtract their numerators, but keep the common denominator the same. Factor the numerator if possible and reduce if there are any factors that can be cancelled.

1. Find the least common denominator for $6a^3b^2$ and $4ab^3$.

These factor into $2 \cdot 3 \cdot a^3 \cdot b^2$ and $2 \cdot 2 \cdot a \cdot b^3$.
The first expression needs to be multiplied by another 2 and b.
The other expression needs to be multiplied by 3 and a^2.
Then both expressions would be $2 \cdot 2 \cdot 3 \cdot a^3 \cdot b^3 = 12a^3b^3 = \text{LCD}$.

2. Find the LCD for $x^2 - 4$, $x^2 + 5x + 6$, and $x^2 + x - 6$.

$$x^2 - 4 \qquad \text{factors into } (x-2)(x+2)$$
$$x^2 + 5x + 6 \quad \text{factors into } (x+3)(x+2)$$
$$x^2 + x - 6 \quad \text{factors into } (x+3)(x-2)$$

To make these lists of factors the same, they must all be $(x+3)(x+2)(x-2)$. This is the LCD.

3.

$$\frac{5}{6a^3b^2} + \frac{1}{4ab^3} = \frac{5(2a)}{6a^3b^2(2a)} + \frac{1(3b^2)}{4ab^3(3b^2)} = \frac{10b}{12a^3b^3} + \frac{3a^2}{12a^3b^3} = \frac{10b + 3a^2}{12a^3b^3}$$

This will not reduce as all 3 terms are not divisible by anything.

4.

$$\frac{2}{x^2-4}-\frac{3}{x^2+5x+6}+\frac{7}{x^2+x-6}=\frac{2}{(x-2)(x+2)}-\frac{3}{(x+3)(x+2)}+$$

$$\frac{7}{(x+3)(x-2)}=\frac{2(x+3)}{(x-2)(x+2)(x+3)}-\frac{3(x-2)}{(x+3)(x+2)(x-2)}+$$

$$\frac{7(x+2)}{(x+3)(x-2)(x+2)}=\frac{2x+6}{(x-2)(x+2)(x+3)}-\frac{3x-6}{(x+3)(x+2)(x-2)}+$$

$$\frac{7x+14}{(x+3)(x-2)(x+2)}=\frac{2x+6-(3x-6)+7x+14}{(x+3)(x-2)(x+2)}=\frac{6x+26}{(x+3)(x-2)(x+2)}$$

This will not reduce.

Try These:

1. $\dfrac{6}{x-3}+\dfrac{2}{x+7}$

2. $\dfrac{5}{4a^2b^5}+\dfrac{3}{5a^4b^3}$

3. $\dfrac{x+3}{x^2-25}+\dfrac{x-6}{x^2-2x-15}$

Rational Expressions – Equivalent Forms

Rational expressions can be changed into other equivalent fractions by either reducing them or by changing them to have a common denominator. When dividing any number of terms by a single term, divide or reduce their coefficients. Then subtract the exponent of a variable on the bottom from the exponent of the same variable from the numerator.

To reduce a rational expression with more than one term in the denominator, the expression must be factored first. Factors that are exactly the same will cancel and each become a 1. Factors that have exactly the opposite signs of each other, such as $(a-b)$ and $(b-a)$, will cancel and one factor becomes a 1 and the other becomes a ⁻1.

To make a fraction have a common denominator, factor the fraction. Determine what factors are missing from that particular denominator, and multiply both the numerator and the denominator by those missing fractions. This gives a new fraction which now has the common denominator.

Simplify these fractions:

1. $\dfrac{24x^3y^6z^3}{8x^2y^2z} = 3xy^4z^2$

2. $\dfrac{3x^2 - 14xy - 5y^2}{x^2 - 25y^2} = \dfrac{(3x+y)(x-5y)}{(x+5y)(x-5y)} = \dfrac{3x+y}{x+5y}$

3. Re-write this fraction with a denominator of $(x+3)(x-5)(x+4)$.

$$\dfrac{x+2}{x^2+7x+12} = \dfrac{x+2}{(x+3)(x+4)} = \dfrac{(x+2)(x-5)}{(x+3)(x+4)(x-5)}$$

Try these:

4. $\dfrac{72x^4y^9z^{10}}{8x^3y^9z^5}$

5. $\dfrac{3x^2 - 13xy - 10y^2}{x^3 - 125y^3}$

6. Re-write this fraction $\dfrac{x+5}{x^2 - 5x - 14}$ with a denominator of $(x+2)(x+3)(x-7)$.

Radical Expressions - Operations

Before you can add or subtract square roots, the numbers or expressions inside the radicals must be the same. First, simplify the radicals, if possible. If the numbers or expressions inside the radicals are the same, add or subtract the numbers (or like expressions) in front of the radicals. Keep the expression inside the radical the same. Be sure that the radicals are as simplified as possible.

Note: If the expressions inside the radicals are not the same, and can not be simplified to become the same, then they can not be combined by addition or subtraction.

To multiply 2 square roots together, follow these steps:

1. Multiply what is outside the radicals together.
2. Multiply what is inside the radicals together.
3. Simplify the radical if possible. Multiply whatever is in front of the radical times the expression that is coming out of the radical.

To divide one square root by another, follow these steps:

1. Work separately on what is inside or outside the square root sign.

2. Divide or reduce the coefficients outside the radical.

3. Divide any like variables outside the radical.

4. Divide or reduce the coefficients inside the radical.

5. Divide any like variables inside the radical.

6. If there is still a radical in the denominator, multiply both the numerator and denominator by the radical in the denominator. Simplify both resulting radicals and reduce again outside the radical (if possible).

Examples:

1. $6\sqrt{7} + 2\sqrt{5} + 3\sqrt{7} = 9\sqrt{7} + 2\sqrt{5}$ These cannot be combined further.

2. $5\sqrt{12} + \sqrt{48} - 2\sqrt{75} = 5\sqrt{2\cdot2\cdot3} + \sqrt{2\cdot2\cdot2\cdot2\cdot3} - 2\sqrt{3\cdot5\cdot5} =$
   $5\cdot2\sqrt{3} + 2\cdot2\sqrt{3} - 2\cdot5\sqrt{3} = 10\sqrt{3} + 4\sqrt{3} - 10\sqrt{3}$ $= 4\sqrt{3}$

3. $(6\sqrt{15x})(7\sqrt{10x}) = 42\sqrt{150x^2} = 42\sqrt{2\cdot3\cdot5\cdot5\cdot x^2} = 42\cdot5x\sqrt{2\cdot3} = 210x\sqrt{6}$

4. $\dfrac{105x^8\sqrt{18x^5y^6}}{30x^2\sqrt{27x^2y^4}} = \dfrac{7x^6\left(x^2\right)\left(y^3\right)\sqrt{2x}}{2(x)\left(y^2\right)\sqrt{3}} = \dfrac{7x^7y\sqrt{2x}}{2\sqrt{3}}$

   $= \dfrac{7x^7y\sqrt{2x}}{2\sqrt{3}}\cdot\dfrac{\sqrt{3}}{\sqrt{3}} = \dfrac{7x^7y\sqrt{6x}}{6}$

Try these:

1. $6\sqrt{24} + 3\sqrt{54} - \sqrt{96}$
2. $\left(2x^2y\sqrt{18x}\right)\left(7xy^7\sqrt{4x}\right)$
3. $\dfrac{125a^5\sqrt{56a^4b^7}}{40a^2\sqrt{40a^2b^8}}$
4. $2\sqrt{3} + 4\sqrt{5} + 6\sqrt{25} - 7\sqrt{9} + 2\sqrt{5} - 8\sqrt{20} - 6\sqrt{16} - 7\sqrt{3}$

Radical Expressions - Simplification

To simplify a radical, follow these steps:

Factor the number or coefficient completely.

For **square roots**, group like factors in pairs. For cube roots, arrange like factors in groups of three. For nth roots, group like factors in groups of n.

For each of these groups, put one of that number outside the radical. Any factors that cannot be combined in groups should be multiplied together and left inside the radical.

The index number of a radical is the little number on the front of the radical. For a cube root, the index is 3. If no index appears, then the index is 2 for square roots.

For variables inside the radical, divide the index number of the radical into each exponent. The quotient (the answer to the division) is the new exponent to be written on the variable outside the radical. The remainder from the division is the new exponent on the variable remaining inside the radical sign. If the remainder is zero, then the variable no longer appears in the radical sign.

Note: Remember that the square root of a negative number can be designated by replacing the negative sign inside that square root with an "i" in front of the radical (to signify an imaginary number). Then simplify the remaining positive radical by the normal method. Include the i outside the radical as part of the answer.

If the index number is an odd number, you can still simplify the radical to get a negative solution.

Example:

$$\sqrt{50a^4b^7} \text{ [SA21]} = \sqrt{5 \cdot 5 \cdot 2 \cdot a^4 \cdot b^7} \text{ [SA22]} = 5a^2b^3\sqrt{2b}$$

Example:

$$7x\sqrt[3]{16x^5} \text{ [SA23]} = 7x\sqrt[3]{2 \cdot 2 \cdot 2 \cdot 2 \cdot x^5} \text{ [SA24]} = 7x \cdot 2x\sqrt[3]{2x^2} \text{ [SA25]} = 14x^2\sqrt[3]{2x^2} \text{ [SA26]}$$

Try These :

1. $\sqrt{72a^9}$

2. $\sqrt{^-98}$

3. $\sqrt[3]{^-8x^6}$

4. $2x^3y\sqrt[4]{243x^6y^{11}}$

Radicals – Multiplication and Division

The conjugate of a binomial is the same expression as the original binomial with the sign between the 2 terms changed.

The conjugate of $3 + 2\sqrt{5}$ is $3 - 2\sqrt{5}$.
The conjugate of $\sqrt{5} - \sqrt{7}$ is $\sqrt{5} + \sqrt{7}$.
The conjugate of $^-6 - \sqrt{11}$ is $^-6 + \sqrt{11}$.

To multiply binomials including radicals, "FOIL" the binomials together. (that is, distribute each term of the first binomial times each term of the second binomial). Multiply what is in front of the radicals together. Multiply what is inside of the two radicals together. Check to see if any of the radicals can be simplified. Combine like terms, if possible.

When one binomial is divided by another binomial, multiply both the numerator and denominator by the conjugate of the denominator. "FOIL" or distribute one binomial through the other binomial. Simplify the radicals, if possible, and combine like terms. Reduce the resulting fraction if every term is divisible outside the radical signs by the same number.

1. $(5+\sqrt{10})(4-3\sqrt{2}) = 20 - 15\sqrt{2} + 4\sqrt{10} - 3\sqrt{20} =$
$20 - 15\sqrt{2} + 4\sqrt{10} - 6\sqrt{5}$

2. $(\sqrt{6}+5\sqrt{2})(3\sqrt{6}-8\sqrt{2}) = 3\sqrt{36} - 8\sqrt{12} + 15\sqrt{12} - 40\sqrt{4} =$
$3\cdot6 - 8\cdot2\sqrt{3} + 15\cdot2\sqrt{3} - 40\cdot2 = 18 - 16\sqrt{3} + 30\sqrt{3} - 80 = {}^-62 + 14\sqrt{3}$

3. $\dfrac{1-\sqrt{2}}{3+5\sqrt{2}} = \dfrac{1-\sqrt{2}}{3+5\sqrt{2}} \cdot \dfrac{3-5\sqrt{2}}{3-5\sqrt{2}} = \dfrac{3-5\sqrt{2}-3\sqrt{2}+5\sqrt{4}}{9-25\sqrt{4}} = \dfrac{3-5\sqrt{2}-3\sqrt{2}+10}{9-50} =$
$\dfrac{13-8\sqrt{2}}{{}^-41}$ or $-\dfrac{13-8\sqrt{2}}{41}$ or $\dfrac{{}^-13+8\sqrt{2}}{41}$

Try These:
1. $(3+2\sqrt{6})(4-\sqrt{6})$
2. $(\sqrt{5}+2\sqrt{15})(\sqrt{3}-\sqrt{15})$
3. $\dfrac{6+2\sqrt{3}}{4-\sqrt{6}}$

Convert Between Rational and Radical Expressions

An expression with a radical sign can be rewritten using a rational exponent. The radicand becomes the base which will have the rational exponent. The index number on the front of the radical sign becomes the denominator of the rational exponent. The numerator of the rational exponent is the exponent which was originally inside the radical sign on the original base. Note: If no index number appears on the front of the radical, then it is a 2. If no exponent appears inside the radical, then use a 1 as the numerator of the rational exponent.

$$\sqrt[5]{b^3} = b^{3/5}$$
$$\sqrt[4]{ab^3} = a^{1/4}b^{3/4}$$

When an expression has a rational exponent, it can be rewritten using a radical sign. The denominator of the rational exponent becomes the index number on the front of the radical sign. The base of the original expression goes inside the radical sign. The numerator of the rational exponent is an exponent which can be placed either inside the radical sign on the original base or outside the radical as an exponent on the radical expression.

$$a^{2/9}b^{4/9}c^{8/9} = \sqrt[9]{a^2 b^4 c^8}$$

$$3^{1/5} = \sqrt[5]{3}$$

If an expression contains rational expressions with different denominators, rewrite the exponents with a common denominator and then change the problem into a radical.

$$a^{2/3}b^{1/2}c^{3/5} = a^{20/30}b^{15/30}c^{18/30} = \sqrt[30]{a^{20}b^{15}c^{18}}$$

Rational Functions

A rational function is given in the form $f(x) = p(x)/q(x)$. In the equation, $p(x)$ and $q(x)$ both represent polynomial functions where $q(x)$ does not equal zero. The branches of rational functions approach asymptotes. Setting the denominator equal to zero and solving will give the value(s) of the vertical asymptotes(s) since the function will be undefined at this point. If the value of $f(x)$ approaches b as the $|x|$ increases, the equation $y = b$ is a horizontal asymptote. To find the horizontal asymptote it is necessary to make a table of values for x that are to the right and left of the vertical asymptotes. The pattern for the horizontal asymptotes will become apparent as the $|x|$ increases.

If there are more than one vertical asymptotes, remember to choose numbers to the right and left of each one in order to find the horizontal asymptotes and have sufficient points to graph the function.

Sample problem:

1. Graph $f(x) = \dfrac{3x+1}{x-2}$.

$x - 2 = 0$
$x = 2$

1. Set denominator $= 0$ to find the vertical asymptote.

x	$f(x)$
3	10
10	3.875
100	3.07
1000	3.007
1	$^-4$
$^-10$	2.417
$^-100$	2.93
$^-1000$	2.99

2. Make table choosing numbers to the right and left of the vertical asymptote.

3. The pattern shows that as the $|x|$ increases $f(x)$ approaches the value 3, therefore a horizontal asymptote exists at $y = 3$

Sketch the graph.

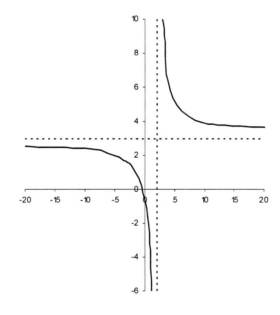

The **absolute value function** for a 1st degree equation is of the form: $y = m(x - h) + k$. Its graph is in the shape of a $\vee$. The point (h,k) is the location of the maximum/minimum point on the graph. "± m" are the slopes of the 2 sides of the $\vee$. The graph opens up if m is positive and down if m is negative.

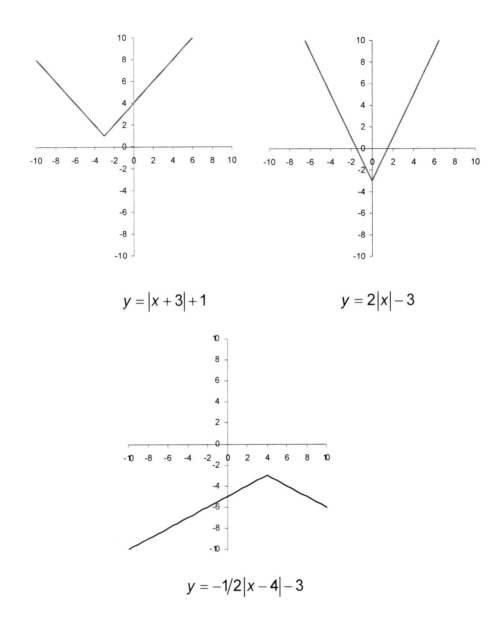

$$y = |x + 3| + 1$$

$$y = 2|x| - 3$$

$$y = -1/2|x - 4| - 3$$

Note that on the first graph, the graph opens up since m is positive 1. It has (–3,1) as its minimum point. The slopes of the two upward rays are ± 1.

The second graph also opens up since m is positive. (0,–3) is its minimum point. The slopes of the 2 upward rays are ± 2.
The third graph is a downward ∧ because m is –1/2. The maximum point on the graph is at (4,–3). The slopes of the two downward rays are ±1/2.

The **identity function** is the linear equation $y = x$. Its graph is a line going through the origin (0,0) and through the first and third quadrants at a $45°$ degree angle.

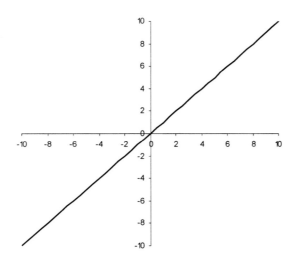

Practice: Graph the following equations:

1. $f(x) = x$
2. $y = -|x - 3| + 5$
3. $y = 3[x]$
4. $y = 2/5|x - 5| - 2$

Rational Equations

Some problems can be solved using equations with rational expressions. First write the equation. To solve it, multiply each term by the LCD of all fractions. This will cancel out all of the denominators and give an equivalent algebraic equation that can be solved.

1. The denominator of a fraction is two less than three times the numerator. If 3 is added to both the numerator and denominator, the new fraction equals $1/2$.

   original fraction: $\dfrac{x}{3x-2}$ revised fraction: $\dfrac{x+3}{3x+1}$

   $$\dfrac{x+3}{3x+1} = \dfrac{1}{2} \qquad\qquad 2x+6 = 3x+1$$

   $$x = 5$$

   original fraction: $\dfrac{5}{13}$

2. Elly Mae can feed the animals in 15 minutes. Jethro can feed them in 10 minutes. How long will it take them if they work together?

   Solution: If Elly Mae can feed the animals in 15 minutes, then she could feed $1/15$ of them in 1 minute, $2/15$ of them in 2 minutes, $x/15$ of them in x minutes. In the same fashion, Jethro could feed $x/10$ of them in x minutes. Together they complete 1 job. The equation is:

   $$\dfrac{x}{15} + \dfrac{x}{10} = 1$$

   Multiply each term by the LCD of 30:

   $$2x + 3x = 30$$
   $$x = 6 \text{ minutes}$$

3. A salesman drove 480 miles from Pittsburgh to Hartford. The next day he returned the same distance to Pittsburgh in half an hour less time than his original trip took, because he increased his average speed by 4 mph. Find his original speed.

   Since distance = rate x time then time = $\dfrac{\text{distance}}{\text{rate}}$

   original time $- 1/2$ hour = shorter return time

   $$\dfrac{480}{x} - \dfrac{1}{2} = \dfrac{480}{x+4}$$

Multiplying by the LCD of $2x(x+4)$, the equation becomes:

$$480\left[2(x+4)\right]-1\left[x(x+4)\right]=480(2x)$$

$$960x+3840-x^2-4x=960x$$

$$x^2+4x-3840=0$$

$(x+64)(x-60)=0$ Either (x-60=0) or (x+64=0) or both=0

$x=60$ 60 mph is the original speed.

$x=64$ This is the solution since the time

                 cannot be negative. Check your answer

$$\frac{480}{60}-\frac{1}{2}=\frac{480}{64}$$

$$8-\frac{1}{2}=7\frac{1}{2}$$

$$7\frac{1}{2}=7\frac{1}{2}$$

Try these: (Check your answers in the Answer Key to Practice Set Problems.)

1. Working together, Larry, Moe, and Curly can paint an elephant in 3 minutes. Working alone, it would take Larry 10 minutes or Moe 6 minutes to paint the elephant. How long would it take Curly to paint the elephant if he worked alone?

2. The denominator of a fraction is 5 more than twice the numerator. If the numerator is doubled, and the denominator is increased by 5, the new fraction is equal to $1/2$. Find the original fraction.

3. A trip from Augusta, Maine to Galveston, Texas is 2108 miles. If one car drove 6 mph faster than a truck and got to Galveston 3 hours before the truck, find the speeds of the car and truck.

- To solve an algebraic formula for some variable, called R, follow the following steps:

a. Eliminate any parentheses using the distributive property.
b. Multiply every term by the LCD of any fractions to write an equivalent equation without any fractions.
c. Move all terms containing the variable, R, to one side of the equation. Move all terms without the variable to the opposite side of the equation.
d. If there are 2 or more terms containing the variable R, factor **only R** out of each of those terms as a common factor.
e. Divide both sides of the equation by the number or expression being multiplied times the variable, R.
 f. Reduce fractions if possible.
g. Remember there are restrictions on values allowed for variables because the denominator can not equal zero.

1. Solve $A = p + prt$ for t.

$$A - p = prt$$

$$\frac{A - p}{pr} = \frac{prt}{pr}$$

$$\frac{A - p}{pr} = t$$

2. Solve $A = p + prt$ for p.

$$A = p(1 + rt)$$

$$\frac{A}{1 + rt} = \frac{p(1 + rt)}{1 + rt}$$

$$\frac{A}{1 + rt} = p$$

3. $A = 1/2 \ h(b_1 + b_2)$ for b_2

$$A = 1/2 \ hb_1 + 1/2 \ hb_2 \quad \leftarrow \text{step a}$$

$$2A = hb_1 + hb_2 \qquad \leftarrow \text{step b}$$

$$2A - hb_1 = hb_2 \qquad \leftarrow \text{step c}$$

$$\frac{2A - hb_1}{h} = \frac{hb_2}{h} \qquad \leftarrow \text{step d}$$

$$\frac{2A - hb_1}{h} = b_2 \qquad \leftarrow \text{will not reduce}$$

Solve:

1. $F = 9/5 \ C + 32$ for C

2. $A = 1/2 \ bh + h^2$ for b

3. $S = 180(n - 2)$ for n

Radical Equations

To solve an **equation with rational expressions**, find the least common denominator of all the fractions. Multiply each term by the LCD of all fractions. This will cancel out all of the denominators and give an equivalent algebraic equation that can be solved. Solve the resulting equation. Once you have found the answer(s), substitute them back into the original equation to check them. Sometimes there are solutions that do not check in the original equation. These are extraneous solutions, which are not correct and must be eliminated. If a problem has more than one potential solution, each solution must be checked separately.

> **NOTE: What this really means is that you can simply just substitute the answers from any multiple choice test back into the question to determine which answer choice is correct.**

Solve and **check**:

1. $\dfrac{72}{x+3} = \dfrac{32}{x+3} + 5$ LCD $= x+3$, so multiply by this.

$$(x+3) \times \frac{72}{x+3} = (x+3) \times \frac{32}{x+3} + 5(x+3)$$

$$72 = 32 + 5(x+3) \rightarrow 72 = 32 + 5x + 15$$

$$72 = 47 + 5x \qquad \rightarrow 25 = 5x$$

$5 = x$ (This checks too).

2. $\dfrac{12}{2x^2 - 4x} + \dfrac{13}{5} = \dfrac{9}{x-2}$ Factor $2x^2 - 4x = 2x(x-2)$.

                               LCD $= 5 \times 2x(x-2)$ or $10x(x-2)$

$$10x(x-2) \times \frac{12}{2x(x-2)} + 10x(x-2) \times \frac{13}{5} = \frac{9}{x-2} \times 10x(x-2)$$

$$60 + 2x(x-2)(13) = 90x$$

$$26x^2 - 142x + 60 = 0$$

$$2(13x^2 - 71x + 30) = 0$$

$2(x-5)(13x-6)$ so $x = 5$ or $x = 6/13$ $\leftarrow$ both check

Try these:

1. $\dfrac{x+5}{3x-5} + \dfrac{x-3}{2x+2} = 1$ 2. $\dfrac{2x-7}{2x+5} = \dfrac{x-6}{x+8}$

To solve a radical equation, follow these steps:

1. Get a radical alone on one side of the equation.
2. Raise both **sides** of the equation to the power equal to the index number. **Do not raise them to that power term by term, but raise the entire side to that power**. Combine any like terms.
3. If there is another radical still in the equation, repeat steps one and two (i.e. get that radical alone on one side of the equation and raise both sides to a power equal to the index). Repeat as necessary until the radicals are all gone.
4. Solve the resulting equation.
5. Once you have found the answer(s), substitute them back into the original equation to check them.

Sometimes there are solutions that do not check in the original equation. These are extraneous solutions, which are not correct and must be eliminated. If a problem has more than one potential solution, each solution must be checked separately.

> **NOTE: What this really means is that you can substitute the answers from any multiple choice test back into the question to determine which answer choice is correct.**

Solve and **check**.

1. $\sqrt{2x+1}+7 = x$

$\sqrt{2x+1} = x-7$

$\left(\sqrt{2x+1}\right)^2 = (x-7)^2$ ← BOTH sides are squared.

$2x+1 = x^2 -14x+49$

$0 = x^2 -16x+48$

$0 = (x-12)(x-4)$

$x = 12, \ x = 4$

When you check these answers in the original equation, 12 checks; however, **4 does not check in the original equation**. Therefore, the only answer is x = 12.

2. $\sqrt{3x+4} = 2\sqrt{x-4}$

$\left(\sqrt{3x+4}\right)^2 = \left(2\sqrt{x-4}\right)^2$

$3x+4 = 4(x-4)$

$3x+4 = 4x-16$

$20 = x$ ← This checks in the original equaion.

3. $\sqrt[4]{7x-3} = 3$

$\left(\sqrt[4]{7x-3}\right)^4 = 3^4$

$7x - 3 = 81$

$7x = 84$

$x = 12$ ← This checks out with the original equation.

4. $\sqrt{x} = {}^-3$

$\left(\sqrt{x}\right)^2 = \left({}^-3\right)^2$

$x = 9$ ← This does NOT check in the original equation.
Since there is no other answer to check,
the correct answer is the empty set or the
null set or $\varnothing$.

Try these:

Solve and check.

1. $\sqrt{8x-24} + 14 = 2x$
2. $\sqrt{6x-2} = 5\sqrt{x-13}$

Absolute Value Equations

To solve an absolute value equation, follow these steps:

1. Get the absolute value expression alone on one side of the equation.

2. Split the absolute value equation into 2 separate equations without absolute value bars. Write the expression inside the absolute value bars (without the bars) equal to the expression on the other side of the equation. Now write the expression inside the absolute value bars equal to the opposite of the expression on the other side of the equation.

3. Now solve each of these equations.

4. **Check each answer by substituting it into the original equation** (with the absolute value symbol). There will be answers that do not check in the original equation. These answers are discarded as they are **extraneous solutions**. If all answers are discarded as incorrect, then the answer to the equation is $\varnothing$, which means the empty set or the null set. (0, 1, or 2 solutions could be correct.)

To solve an absolute value inequality, follow these steps:

1. Get the absolute value expression alone on one side of the inequality. Remember: **Dividing or multiplying by a negative number will reverse the direction of the inequality sign.**

2. Remember what the inequality sign is at this point.

3. Split the absolute value inequality into 2 separate inequalities without absolute value bars. First rewrite the inequality without the absolute bars and solve it. Next write the expression inside the absolute value bar followed by the opposite inequality sign and then by the opposite of the expression on the other side of the inequality. Now solve it.

4. If the sign in the inequality on step 2 is $<$ or $\leq$, the answer is those 2 inequalities connected by the word **and**. The solution set consists of the points between the 2 numbers on the number line. If the sign in the inequality on step 2 is $>$ or $\geq$, the answer is those 2 inequalities connected by the word **or**. The solution set consists of the points outside the 2 numbers on the number line.

If an expression inside an absolute value bar is compared to a negative number, the answer can also be either all real numbers or the empty set ($\varnothing$). For instance,

$$|x+3| < {}^{-}6$$

would have the empty set as the answer, since an absolute value is always positive and will never be less than $^{-}6$. However,

$$|x+3| > {}^{-}6$$

would have all real numbers as the answer, since an absolute value is always positive or at least zero, and will never be less than -6. In similar fashion,

$$|x+3| = {}^{-}6$$

would never check because an absolute value will never give a negative value.

Example: Solve and check:
$$|2x-5| + 1 = 12$$
$$|2x-5| = 11 \qquad \text{Get absolute value alone.}$$

Rewrite as 2 equations and solve separately.

same equation without absolute value		same equation without absolute value but right side is opposite
$2x-5 = 11$		$2x-5 = {}^{-}11$
$2x = 16$	and	$2x = {}^{-}6$
$x = 8$		$x = {}^{-}3$

Checks:	$|2x-5| + 1 = 12$	$|2x-5| + 1 = 12$
	$|2(8)-5| + 1 = 12$	$|2({}^{-}3)-5| + 1 = 12$
	$|11| + 1 = 12$	$|{}^{-}11| + 1 = 12$
	$12 = 12$	$12 = 12$

This time both 8 and $^{-}3$ check.

Example: Solve and check:

$$2|x - 7| - 13 \geq 11$$

$$2|x - 7| \geq 24 \qquad \text{Get absolute value alone.}$$

$$|x - 7| \geq 12$$

Rewrite as 2 inequalities and solve separately.

same inequality without absolute value		same inequality without absolute value but right side and inequality sign are both the opposite
$x - 7 \geq 12$	or	$x - 7 \leq {}^-12$
$x \geq 19$	or	$x \leq {}^-5$

COMPETENCY 0010 **UNDERSTAND THE PRINCIPLES AND PROPERTIES OF EXPONENTIAL AND LOGARITHMIC FUNCTIONS.**

The basic **exponential function** is defined by $f(x) = B^x$. Stated another way, it is the function consisting of the base of the natural logarithm e taken to the power of the variable. e is the constant 2.718....

When changing common logarithms to exponential form,

$$y = \log_b x \quad \text{if and only if} \quad x = b^y$$

Natural logarithms can be changed to exponential form by using,

$$\log_e x = \ln x \quad \text{or} \quad \ln x = y \quad \text{can be written as } e^y = x$$

Practice Problems:

Express in exponential form.

1. $\log_3 81 = 4$

    $x = 81 \quad b = 3 \quad y = 4$ Identify values.

    $81 = 3^4$ Rewrite in exponential form.

Solve by writing in exponential form.

2. $\log_x 125 = 3$

    $x^3 = 125$ Write in exponential form.

    $x^3 = 5^3$ Write 125 in exponential form.

    $x = 5$ Bases must be equal if exponents are equal.

Use a scientific calculator to solve.

3. Find $\ln 72$.

    $\ln 72 = 4.2767$ Use the $\ln x$ key to find natural logs.

4. Find $\ln x = 4.2767$ Write in exponential form.

    $e^{4.2767} = x$ Use the key (or 2nd $\ln x$) to find z.

    $x = 72.002439$ The small difference is due to rounding.

A **logarithmic regression** equation is of the form: $y = a + b\ln x$.

Example:

The water, w, in an open container is evaporating. The number of ounces remaining after h hours is shown in the table below:

Hours (h)	2	5	10	15	19	30
Water (w)	13	11	9	8.5	7.5	6.5

We construct a scatter plot for this data and find a logarithmic regression equation to model the data.

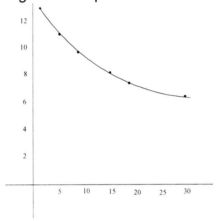

$$w = 14.71 - 2.41 \ln h$$

Using this regression equation predict how many ounces of water are remaining in the container after 48 hours.

$$w = 14.71 - 2.41\ln(48)$$
$$w = 14.71 - 2.41(3.87)$$
$$w = 14.71 - 9.33$$
$$w = 5.4$$

Power regression is a functional form used for nonlinear regression. The formula is: $Y = ab^X$.

Example:

On a newly discovered planet the weight, w, of an object and the distance, d, this object is from the surface of the planet were recorded and are shown in the table below (Weight is in pounds and distance in miles).

d	w
2	116
3	58
4	36
5	24
6	18

We construct a scatter plot for the given data using distance as the independent variable (x) and find a power regression equation to model this data.

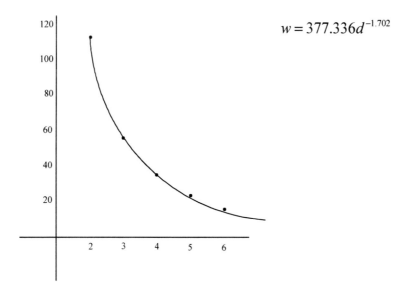

$$w = 377.336d^{-1.702}$$

Using the regression equation we find the predicted weight of the object to the nearest pound when it is 20 miles from the surface of the planet.

$$w = 377.336(20)^{-1.702}$$
$$= 377.336(.006)$$
$$= 2$$

To solve logarithms or exponential functions it is necessary to use several properties.

Multiplication Property
$$\log_b mn = \log_b m + \log_b n$$

Quotient Property
$$\log_b \frac{m}{n} = \log_b m - \log_b n$$

Powers Property
$$\log_b n^r = r \log_b n$$

Equality Property
$$\log_b n = \log_b m \qquad \text{if and only if } n = m.$$

Change of Base Formula
$$\log_b n = \frac{\log n}{\log b}$$

$$\log_b b^x = x \text{ and } b^{\log_b x} = x$$

Sample problem.

Solve for x.

1. $\log_6(x-5) + \log_6 x = 2$

   $\log_6 x(x-5) = 2$ Use product property.

   $\log_6 x^2 - 5x = 2$ Distribute.

   $x^2 - 5x = 6^2$ Write in exponential form.

   $x^2 - 5x - 36 = 0$ Solve quadratic equation.

   $(x+4)(x-9) = 0$

     $x = {}^-4$ $x = 9$

***Be sure to check results. Remember x must be greater than zero in $\log x = y$.

Check: $\log_6(x-5) + \log_6 x = 2$

   $\log_6({}^-4 - 5) + \log_6({}^-4) = 2$ Substitute the first answer ${}^-4$.

   $\log_6({}^-9) + \log_6({}^-4) = 2$ This is undefined, x is less than zero.

   $\log_6(9-5) + \log_6 9 = 2$ Substitute the second answer 9.

   $\log_6 4 + \log_6 9 = 2$

   $\log_6(4)(9) = 2$ Multiplication property.

   $\log_6 36 = 2$

   $6^2 = 36$ Write in exponential form.

   $36 = 36$

Practice problems:

1. $\log_4 x = 2\log_4 3$

2. $2\log_3 x = 2 + \log_3(x-2)$

3. Use change of base formula to find $(\log_3 4)(\log_4 3)$.

SUBAREA III. **MEASUREMENT AND GEOMETRY**

COMPETENCY 0011 **UNDERSTAND PRINCIPLES AND PROCEDURES RELATED TO MEASUREMENT.**

FIGURE	AREA FORMULA	PERIMETER FORMULA
Rectangle	LW	$2(L+W)$
Triangle	$\frac{1}{2}bh$	$a+b+c$
Parallelogram	bh	sum of lengths of sides
Trapezoid	$\frac{1}{2}h(a+b)$	sum of lengths of sides

Sample problems:

1. Find the area and perimeter of a rectangle if its length is 12 inches and its diagonal is 15 inches.

    1. Draw and label a sketch.

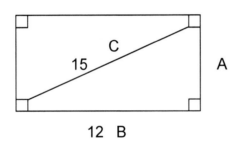

2. Since the height is still needed use Pythagorean formula to find missing leg of the triangle.

$$A^2 + B^2 = C^2$$
$$A^2 + 12^2 = 15^2$$
$$A^2 = 15^2 - 12^2$$
$$A^2 = 81$$
$$A = 9$$

Now use this information to find the area and perimeter.

$A = LW$	$P = 2(L+W)$	1. write formula
$A = (12)(9)$	$P = 2(12+9)$	2. substitute
$A = 108\ \text{in}^2$	$P = 42$ inches	3. solve

Circles – Area and Perimeter

Given a circular figure the formulas are as follows:

$$A = \pi r^2 \qquad\qquad C = \pi d \quad \text{or} \quad 2\pi r$$

Sample problem:

1. If the area of a circle is 50 cm^2, find the circumference.

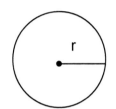

$A = 50 \text{ cm}^2$

1. Draw a sketch.

2. Determine what is still needed.

Use the area formula to find the radius.

$A = \pi r^2$	1. write formula
$50 = \pi r^2$	2. substitute
$\dfrac{50}{\pi} = r^2$	3. divide by π
$15.915 = r^2$	4. substitute
$\sqrt{15.915} = \sqrt{r^2}$	5. take square root of both sides
$3.989 \approx r$	6. compute

Use the approximate answer (due to rounding) to find the circumference.

$C = 2\pi r$	1. write formula
$C = 2\pi\,(3.989)$	2. substitute
$C \approx 25.064$	3. compute

Composite Figures - Area

Use appropriate problem solving strategies to find the solution.

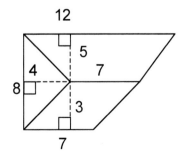

1. Find the area of the given figure.

2. Cut the figure into familiar shapes.

3. Identify what type figures are given and write the appropriate formulas.

Area of figure 1 (triangle)	Area of figure 2 (parallelogram)	Area of figure 3 (trapezoid)
$A = \dfrac{1}{2}bh$	$A = bh$	$A = \dfrac{1}{2}h(a+b)$
$A = \dfrac{1}{2}(8)(4)$	$A = (7)(3)$	$A = \dfrac{1}{2}(5)(12+7)$
$A = 16$ sq. ft	$A = 21$ sq. ft	$A = 47.5$ sq. ft

Now find the total area by adding the area of all figures.

Total area $= 16 + 21 + 47.5$
Total area $= 84.5$ square ft

Regular Polygons

Given the figure below, find the area by dividing the polygon into smaller shapes.

1. divide the figure into two triangles and a rectangle.

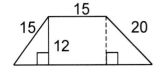

2. find the missing lengths.

3. find the area of each part.

4. find the sum of all areas.

Find base of both right triangles using Pythagorean Formula:

$$a^2 + b^2 = c^2 \qquad\qquad a^2 + b^2 = c^2$$
$$a^2 + 12^2 = 15^2 \qquad\qquad a^2 + 12^2 = 20^2$$
$$a^2 = 225 - 144 \qquad\qquad a^2 = 400 - 144$$
$$a^2 = 81 \qquad\qquad\qquad a^2 = 256$$
$$a = 9 \qquad\qquad\qquad\quad a = 16$$

Area of triangle 1	Area of triangle 2	Area of rectangle
$A = \dfrac{1}{2}bh$	$A = \dfrac{1}{2}bh$	$A = LW$
$A = \dfrac{1}{2}(9)(12)$	$A = \dfrac{1}{2}(16)(12)$	$A = (15)(12)$
$A = 54$ sq. units	$A = 96$ sq. units	$A = 180$ sq. units

Find the sum of all three figures.

$54 + 96 + 180 = 330$ square units

Solid Figures

Use the formulas to find the volume and surface area.

FIGURE	VOLUME	TOTAL SURFACE AREA
Right Cylinder	$\pi r^2 h$	$2\pi rh + 2\pi r^2$
Right Cone	$\dfrac{\pi r^2 h}{3}$	$\pi r\sqrt{r^2 + h^2} + \pi r^2$
Sphere	$\dfrac{4}{3}\pi r^3$	$4\pi r^2$
Rectangular Solid	LWH	$2LW + 2WH + 2LH$

Note: $\sqrt{r^2 + h^2}$ is equal to the slant height of the cone.

Sample problem:

1. Given the figure below, find the volume and surface area.

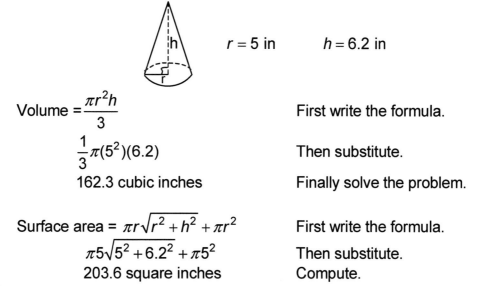

$r = 5$ in $h = 6.2$ in

Volume $= \dfrac{\pi r^2 h}{3}$ First write the formula.

$\dfrac{1}{3}\pi(5^2)(6.2)$ Then substitute.

162.3 cubic inches   Finally solve the problem.

Surface area $= \pi r\sqrt{r^2 + h^2} + \pi r^2$ First write the formula.

$\pi 5\sqrt{5^2 + 6.2^2} + \pi 5^2$ Then substitute.

203.6 square inches   Compute.

Note: volume is always given in cubic units and area is always given in square units.

Right Prisms and Regular Pyramids – Area and Volume

FIGURE	LATERAL AREA	TOTAL AREA	VOLUME
Right prism	sum of area of lateral faces (rectangles)	lateral area plus 2 times the area of base	area of base times height
regular pyramid	sum of area of lateral faces (triangles)	lateral area plus area of base	1/3 times the area of the base times the height

Find the total area of the given figure:

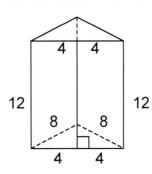

1. Since this is a triangular prism, first find the area of the bases.

2. Find the area of each rectangular lateral face.

3. Add the areas together.

$A = \dfrac{1}{2}bh$ $A = LW$ 1. write formula

$8^2 = 4^2 + h^2$
$h = 6.928$ 2. find the height of the base triangle

$A = \dfrac{1}{2}(8)(6.928)$ $A = (8)(12)$

        3. substitute known values

$A = 27.713$ sq. units $A = 96$ sq. units   4. compute

Total Area $= 2(27.713) + 3(96)$
$= 343.426$ sq. units

Right Circular Cylinders and Cones – Area and Volume

FIGURE	VOLUME	TOTAL SURFACE AREA	LATERAL AREA
Right Cylinder	$\pi r^2 h$	$2\pi rh + 2\pi r^2$	$2\pi rh$
Right Cone	$\dfrac{\pi r^2 h}{3}$	$\pi r\sqrt{r^2 + h^2} + \pi r^2$	$\pi r\sqrt{r^2 + h^2}$

Note: $\sqrt{r^2 + h^2}$ is equal to the slant height of the cone.

Sample problem:

1. A water company is trying to decide whether to use traditional cylindrical paper cups or to offer conical paper cups since both cost the same. The traditional cups are 8 cm wide and 14 cm high. The conical cups are 12 cm wide and 19 cm high. The company will use the cup that holds the most water.

1. Draw and label a sketch of each.

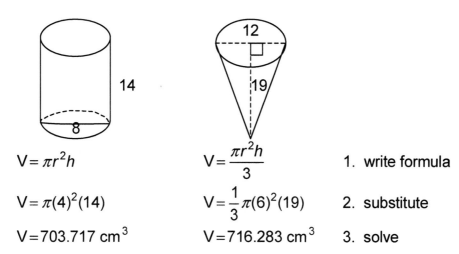

$V = \pi r^2 h$ $V = \dfrac{\pi r^2 h}{3}$ 1. write formula

$V = \pi (4)^2 (14)$ $V = \dfrac{1}{3} \pi (6)^2 (19)$ 2. substitute

$V = 703.717 \text{ cm}^3$ $V = 716.283 \text{ cm}^3$ 3. solve

The choice should be the conical cup since its volume is more.

Spheres – Surface Area and Volume

FIGURE	VOLUME	TOTAL SURFACE AREA
Sphere	$\dfrac{4}{3} \pi r^3$	$4 \pi r^2$

Sample problem:

1. How much material is needed to make a basketball that has a diameter of 15 inches? How much air is needed to fill the basketball?

Draw and label a sketch:

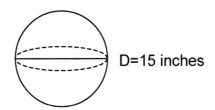 D=15 inches

Total surface area	Volume	
$TSA = 4\pi r^2$	$V = \frac{4}{3}\pi r^3$	1. write formula
$= 4\pi(7.5)^2$	$= \frac{4}{3}\pi(7.5)^3$	2. substitute
$= 706.9$ in^2	$= 1767.1$ in^3	3. solve

Changes in Area – Two-Dimensional Figures

The strategy for solving problems of this nature should be to identify the given shapes and choose the correct formulas. Subtract the smaller cut out shape from the larger shape.

Sample problems:

1. Find the area of one side of the metal in the circular flat washer shown below:

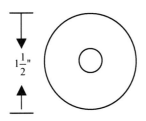

1. the shapes are both circles.

2. use the formula A = πr^2 for both.

   (Inside diameter is 3/8")

Area of larger circle	Area of smaller circle
A = πr^2	A = πr^2
A = $\pi(.75^2)$	A = $\pi(.1875^2)$
A = 1.76625 in^2	A = .1103906 in^2 [SA27]

Area of metal washer = larger area - smaller area

$$= 1.76625 \text{ in}^2 - .1103906 \text{ in}^2 \text{ [SA28]}$$

$$= 1.6558594 \text{ in}^2 \text{ [SA29]}$$

2. You have decided to fertilize your lawn. The shapes and dimensions of your lot, house, pool and garden are given in the diagram below. The shaded area will not be fertilized. If each bag of fertilizer costs $7.95 and covers 4,500 square feet, find the total number of bags needed and the total cost of the fertilizer.

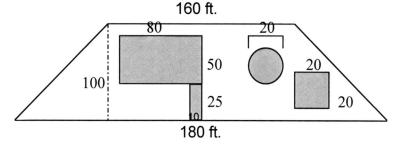

Area of Lot

$A = \frac{1}{2}\, h(b_1 + b_2)$

$A = \frac{1}{2}(100)(180 + 160)$

$A = 17,000$ sq ft

Area of House

$A = LW$

$A = (80)(50)$

$A = 4,000$ sq ft

Area of Driveway

$A = LW$

$A = (10)(25)$

$A = 250$ sq ft

Area of Pool

$A = \pi r^2$

$A = \pi (10)^2$

$A = 314.159$ sq. ft.

Area of Garden

$A = s^2$

$A = (20)^2$

$A = 400$ sq. ft.

Total area to fertilize = Lot area - (House + Driveway + Pool + Garden)

= 17,000 - (4,000 + 250 + 314.159 + 400)
= 12,035.841 sq ft

Number of bags needed = Total area to fertilize / 4,500 sq.ft. bag

= 12,035.841 / 4,500

= 2.67 bags

Since we cannot purchase 2.67 bags we must purchase 3 full bags.

Total cost = Number of bags * $7.95
= 3 * $7.95
= $23.85

Examining the change in area or volume of a given figure requires first to find the existing area given the original dimensions and then finding the new area given the increased dimensions.

Sample problem:

Given the rectangle below determine the change in area if the length is increased by 5 and the width is increased by 7.

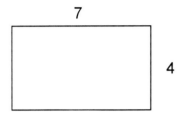

Draw and label a sketch of the new rectangle.

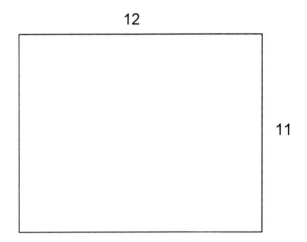

Find the areas.

Area of original = LW   Area of enlarged shape = LW
          = (7)(4)   = (12)(11)
          = 28 units^2 = 132 units^2

The change in area is 132 – 28 = 104 units^2.

Compound Shapes - Area

Cut the compound shape into smaller, more familiar shapes and then compute the total area by adding the areas of the smaller parts.

Sample problem:

Find the area of the given shape.

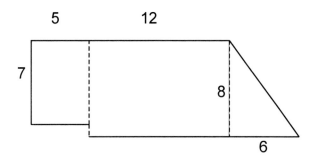

1. Using a dotted line we have cut the shape into smaller parts that are familiar.

2. Use the appropriate formula for each shape and find the sum of all areas.

Area 1 = LW	Area 2 = LW	Area 3 = ½bh
= (5)(7)	= (12)(8)	= ½(6)(8)
= 35 units^2	= 96 units^2	= 24 units^2

Total area = Area 1 + Area 2 + Area 3
= 35 + 96 + 24
= 155 units^2

Triangles – Exterior and Interior Angles

The sum of the measures of the angles of a triangle is $180°$.

Example 1:
Can a triangle have two right angles?
No. A right angle measures $90°$, therefore the sum of two right angles would be $180°$ and there could not be third angle.

Example 2:
Can a triangle have two obtuse angles?
No. Since an obtuse angle measures more than $90°$ the sum of two obtuse angles would be greater than $180°$.

Example 3:
Can a right triangle be obtuse?
No. Once again, the sum of the angles would be more than $180°$.

Example 4:
In a triangle, the measure of the second angle is three times the first. The third angle equals the sum of the measures of the first two angles. Find the number of degrees in each angle.

Let x = the number of degrees in the first angle
$3x$ = the number of degrees in the second angle
$x + 3x$ = the measure of the third angle

Since the sum of the measures of all three angles is 180°.

$$x + 3x + (x + 3x) = 180$$
$$8x = 180$$
$$x = 22.5$$
$$3x = 67.5$$
$$x + 3x = 90$$

Thus the angles measure 22.5°, 67.5°, and 90°. Additionally, the triangle is a right triangle.

Exterior Angles

Two adjacent angles form a linear pair when they have a common side and their remaining sides form a straight angle. Angles in a linear pair are supplementary. An exterior angle of a triangle forms a linear pair with an angle of the triangle.

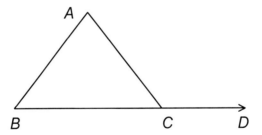

$\angle ACD$ is an exterior angle of triangle ABC, forming a linear pair with $\angle ACB$.

The measure of an exterior angle of a triangle is equal to the sum of the measures of the two non-adjacent interior angles.

Example:
In triangle *ABC*, the measure of $\angle A$ is twice the measure of $\angle B$. $\angle C$ is 30° more than their sum. Find the measure of the exterior angle formed at $\angle C$.

$\begin{aligned}
\text{Let}\quad x &= \text{the measure of } \angle B \\
2x &= \text{the measure of } \angle A \\
x + 2x + 30 &= \text{the measure of } \angle C \\
x + 2x + x + 2x + 30 &= 180 \\
6x + 30 &= 180 \\
6x &= 150 \\
x &= 25 \\
2x &= 50
\end{aligned}$

It is not necessary to find the measure of the third angle, since the exterior angle equals the sum of the opposite interior angles. Thus the exterior angle at $\angle C$ measures 75°.

Convex Polygons – Exterior and Interior Angles

A **quadrilateral** is a polygon with four sides.
The sum of the measures of the angles of a quadrilateral is 360°.

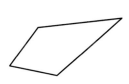 

A **trapezoid** is a quadrilateral with exactly <u>one</u> pair of parallel sides.

In an **isosceles trapezoid**, the non-parallel sides are congruent.

A **parallelogram** is a quadrilateral with <u>two</u> pairs of parallel sides.

A **rectangle** is a parallelogram with a right angle.

A **rhombus** is a parallelogram with all sides of equal length.

A **square** is a rectangle with all sides of equal length.

Regular Polygons – Exterior and Interior Angles

A **polygon** is a simple closed figure composed of line segments.

In a **regular polygon** all sides are the same length and all angles are the same measure.

The sum of the measures of the **interior angles** of a polygon can be determined using the following formula, where n represents the number of angles in the polygon.

Sum of $\angle s = 180(n - 2)$

The measure of each angle of a regular polygon can be found by dividing the sum of the measures by the number of angles.

Measure of $\angle = \dfrac{180(n-2)}{n}$

Example: Find the measure of each angle of a regular octagon. Since an octagon has eight sides, each angle equals:

$$\frac{180(8-2)}{8} = \frac{180(6)}{8} = 135^{\circ}$$

The sum of the measures of the **exterior angles** of a polygon, taken one angle at each vertex, equals 360°.

The measure of each exterior angle of a regular polygon can be determined using the following formula, where n represents the number of angles in the polygon.

Measure of exterior $\angle$ of regular polygon = $180 - \dfrac{180(n-2)}{n}$

or, more simply = $\dfrac{360}{n}$

Example: Find the measure of the interior and exterior angles of a regular pentagon.

Since a pentagon has five sides, each exterior angle measures:

$$\dfrac{360}{5} = 72^{\circ}$$

Since each exterior angle is supplementary to its interior angle, the interior angle measures 180 - 72 or 108°.

Trapezoids

A **trapezoid** is a quadrilateral with exactly <u>one</u> pair of parallel sides.

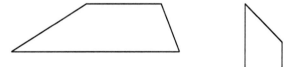

In an **isosceles trapezoid**, the non-parallel sides are congruent.

An isosceles trapezoid has the following properties:

The diagonals of an isosceles trapezoid are congruent.
The base angles of an isosceles trapezoid are congruent.

Example:

An isosceles trapezoid has a diagonal of 10 and a base angle measure of $30°$. Find the measure of the other 3 angles.

Based on the properties of trapezoids, the measure of the other base angle is $30°$ and the measure of the other diagonal is 10.

The other two angles have a measure of:

$$360 = 30(2) + 2x$$

$x = 150°$ The other two angles measure $150°$ each.

Finding the **rate of change** of one quantity (for example distance, volume, etc.) with respect to time it is often referred to as a rate of change problem. To find an instantaneous rate of change of a particular quantity, write a function in terms of time for that quantity; then take the derivative of the function. Substitute in the values at which the instantaneous rate of change is sought.

Functions which are in terms of more than one variable may be used to find related rates of change. These functions are often not written in terms of time. To find a related rate of change, follow these steps.

1. Write an equation which relates all the quantities referred to in the problem.

2. Take the derivative of both sides of the equation with respect to time. Follow the same steps as used in implicit differentiation. This means take the derivative of each part of the equation remembering to multiply each term by the derivative of the variable involved with respect to time. For example, if a term includes the variable v for volume, take the derivative of the term remembering to multiply by dv/dt for the derivative of volume with respect to time. dv/dt is the rate of change of the volume.

3. Substitute the known rates of change and quantities, and solve for the desired rate of change.

Example:

1. What is the instantaneous rate of change of the area of a circle where the radius is 3 cm?

$A(r) = \pi r^2$	Write an equation for area.
$A'(r) = 2\pi r$	Take the derivative to find the rate of change.
$A'(3) = 2\pi(3) = 6\pi$	Substitute in $r = 3$ to arrive at the instantaneous rate of change.

There are many methods for converting measurements within a system. One method is to multiply the given measurement by a conversion factor. This conversion factor is the ratio of:

$$\frac{\text{new units}}{\text{old units}} \quad \text{OR} \quad \frac{\text{what you want}}{\text{what you have}}$$

Sample problems:

1. Convert 3 miles to yards.

$$\frac{3 \text{ miles}}{1} \times \frac{1{,}760 \text{ yards}}{1 \text{ mile}} = \frac{\quad}{\quad} \text{yards}$$

1. multiply by the conversion factor

$$= 5{,}280 \text{ yards}$$

2. cancel the miles units

3. solve

2. Convert 8,750 meters to kilometers.

$$\frac{8{,}750 \text{ meters}}{1} \times \frac{1 \text{ kilometer}}{1000 \text{ meters}} = \frac{\quad}{\quad} \text{km}$$

1. multiply by the conversion factor

$$= 8.75 \text{ kilometers}$$

2. cancel the meters units

3. solve

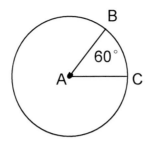

Central angle BAC $= 60°$

Minor arc BC $= 60°$

Major arc BC $= 360 - 60 = 300°$

If you draw two radii in a circle, the angle they form with the center as the vertex is a central angle. The piece of the circle "inside" the angle is an arc. Just like a central angle, an arc can have any degree measure from 0 to 360. The measure of an arc is equal to the measure of the central angle which forms the arc. Since a diameter forms a semicircle and the measure of a straight angle like a diameter is 180°, the measure of a semicircle is also 180°.

Given two points on a circle, there are two different arcs which the two points form. Except in the case of semicircles, one of the two arcs will always be greater than 180° and the other will be less than 180°. The arc less than 180° is a minor arc and the arc greater than 180° is a major arc.

Examples:

1.

$m\square BAD = 45°$

What is the measure of the major arc BD?

$\square BAD = $ minor arc BD

$45° = $ minor arc BD

$360 - 45 = $ major arc BD

$315° = $ major arc BD

The measure of the central angle is the same as the measure of the arc it forms. A major and minor arc always add to $360°$.

2.

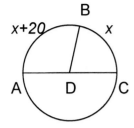

$\overline{AC}$ is a diameter of circle D.
What is the measure of $\angle BDC$?

$m\angle ADB + m\angle BDC = 180°$

$x + 20 + x = 180$

$2x + 20 = 180$

$2x = 160$

$x = 80$

minor arc $BC = 80°$

$m\angle BDC = 80°$

A diameter forms a semicircle
which has a measure of $180°$.

A central angle has the same
measure as the arc it forms.

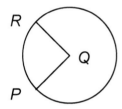

$$\frac{\angle PQR}{360°} = \frac{\text{length of arc } RP}{\text{circumference of } \odot Q} = \frac{\text{area of sector } PQR}{\text{area of } \odot Q}$$

While an arc has a measure associated to the degree measure of a central angle, it also has a length which is a fraction of the circumference of the circle.

For each central angle and its associated arc, there is a sector of the circle which resembles a pie piece. The area of such a sector is a fraction of the area of the circle.

The fractions used for the area of a sector and length of its associated arc are both equal to the ratio of the central angle to $360°$.

Examples:

1.

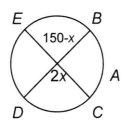

$\square\,A$ has a radius of 4 cm. What is the length of arc ED?

$$2x + 150 - x = 180$$
$$x + 150 = 180$$
$$x = 30$$

Arc BE and arc DE make a semicircle.

Arc $ED = 2(30) = 60°$

The ratio $60°$ to $360°$ is equal to the ratio of arch length ED to the circumference of $\square\,A$.

$$\frac{60}{360} = \frac{\text{arc length } ED}{2\pi 4}$$

$$\frac{1}{6} = \frac{\text{arc length}}{8\pi}$$

$$\frac{8\pi}{6} = \text{arc length}$$

Cross multiply and solve for the arc length.

2.

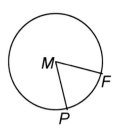

The radius of ⊡ M is 3 cm. The length of arc PF is 2π cm. What is the area of sector PMF?

Circumference of ⊡ $M = 2\pi(3) = 6\pi$

Area of ⊡ $M = \pi(3)^2 = 9\pi$

Find the circumference and area of the circle.

$$\frac{\text{area of } PMF}{9\pi} = \frac{2\pi}{6\pi}$$

The ratio of the sector area to the circle area is the same as the arc length to the circumference.

$$\frac{\text{area of } PMF}{9\pi} = \frac{1}{3}$$

$$\text{area of } PMF = \frac{9\pi}{3}$$

$$\text{area of } PMF = 3\pi$$

Solve for the area of the sector.

Angles with their vertices on the circle:

An inscribed angle is an angle whose vertex is on the circle. Such an angle could be formed by two chords, two diameters, two secants, or a secant and a tangent. An inscribed angle has one arc of the circle in its interior. The measure of the inscribed angle is one-half the measure of this intercepted arc. If two inscribed angles intercept the same arc, the two angles are congruent (i.e. their measures are equal). If an inscribed angle intercepts an entire semicircle, the angle is a right angle.

Angles with their vertices in a circle's interior:

When two chords intersect inside a circle, two sets of vertical angles are formed. Each set of vertical angles intercepts two arcs which are across from each other. The measure of an angle formed by two chords in a circle is equal to one-half the sum of the angle intercepted by the angle and the arc intercepted by its vertical angle.

Angles with their vertices in a circle's exterior:

If an angle has its vertex outside of the circle and each side of the circle intersects the circle, then the angle contains two different arcs. The measure of the angle is equal to one-half the difference of the two arcs.

Examples:

1.

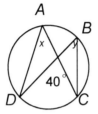

Find *x and y.*
arc $DC = 40°$

$$m\angle DAC = \frac{1}{2}(40) = 20°$$

$\angle DAC$ and $\angle DBC$ are both inscribed angles, so each one has a measure equal to one-half the measure of arc *DC*.

$$m\angle DBC = \frac{1}{2}(40) = 20°$$

$x = 20°$ and $y = 20°$

COMPETENCY 0012 **UNDERSTAND THE PRINCIPLES AND PROPERTIES OF EUCLIDEAN GEOMETRY IN TWO AND THREE DIMENSIONS.**

Angles – Labeling and Classification

The classifying of angles refers to the angle measure. The naming of angles refers to the letters or numbers used to label the angle.

Sample Problem:

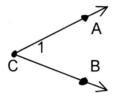

$\overrightarrow{CA}$ (read ray CA) and $\overrightarrow{CB}$ are the sides of the angle.
The angle can be called $\angle ACB$, $\angle BCA$, $\angle C$ or $\angle 1$.

Angles are classified according to their size as follows:

acute:	greater than 0 and less than 90 degrees
right:	exactly 90 degrees
obtuse:	greater than 90 and less than 180 degrees
straight:	exactly 180 degrees

Angles can be classified in a number of ways. Some of those classifications are outlined here.

Adjacent angles have a common vertex and one common side but no interior points in common.

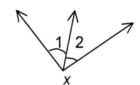

Complementary angles add up to 90 degrees.

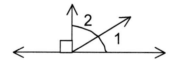

Supplementary angles add up to 180 degrees.

Vertical angles have sides that form two pairs of opposite rays.

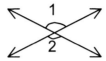

Corresponding angles are in the same corresponding position on two parallel lines cut by a transversal.

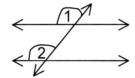

Alternate interior angles are diagonal angles on the inside of two parallel lines cut by a transversal.

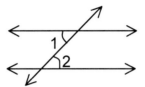

Alternate exterior angles are diagonal angles on the outside of two parallel lines cut by a transversal.

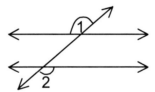

Lines and Planes – Classification

Parallel lines or planes do not intersect.

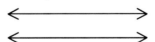

Perpendicular lines or planes form a 90-degree angle to each other.

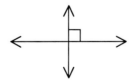

Intersecting lines share a common point and intersecting planes share a common set of points or line.

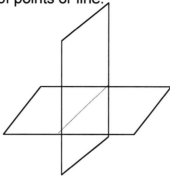

Skew lines do not intersect and do not lie on the same plane.

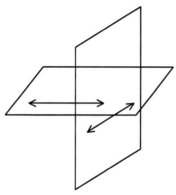

Congruent Polygons

Congruent figures have the same size and shape. If one is placed above the other, it will fit exactly. Congruent lines have the same length. Congruent angles have equal measures. The symbol for congruent is $\cong$.

Polygons (pentagons) *ABCDE* and *VWXYZ* are congruent. They are exactly the same size and shape.

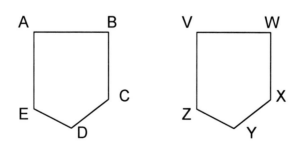

$ABCDE \cong VWXYZ$

Corresponding parts are those congruent angles and congruent sides, that is:

corresponding angles	corresponding sides
$\angle A \leftrightarrow \angle V$	$AB \leftrightarrow VW$
$\angle B \leftrightarrow \angle W$	$BC \leftrightarrow WX$
$\angle C \leftrightarrow \angle X$	$CD \leftrightarrow XY$
$\angle D \leftrightarrow \angle Y$	$DE \leftrightarrow YZ$
$\angle E \leftrightarrow \angle Z$	$AE \leftrightarrow VZ$

Similar Polygons

Two figures that have the **same shape** are **similar**.
Two polygons are similar if corresponding angles are congruent and corresponding sides are in proportion. Corresponding parts of similar polygons are proportional.

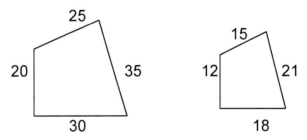

Similar Triangles

AA Similarity Postulate

If two angles of one triangle are congruent to two angles of another triangle, then the triangles are similar.

SAS Similarity Theorem

If an angle of one triangle is congruent to an angle of another triangle and the sides adjacent to those angles are in proportion, then the triangles are similar.

SSS Similarity Theorem

If the sides of two triangles are in proportion, then the triangles are similar.

Problem Solving

Example:

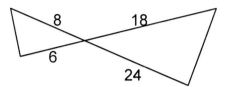

The two triangles are similar since the sides are proportional and vertical angles are congruent.

Example: Given two similar quadrilaterals. Find the lengths of sides x, y, and z.

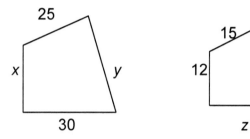

Since corresponding sides are proportional:

$$\frac{15}{25} = \frac{3}{5} \text{ so the scale is } \frac{3}{5}$$

$$\frac{12}{x} = \frac{3}{5} \qquad \frac{21}{y} = \frac{3}{5} \qquad \frac{z}{30} = \frac{3}{5}$$

$$3x = 60 \qquad 3y = 105 \qquad 5z = 90$$
$$x = 20 \qquad y = 35 \qquad z = 18$$

Polygons are similar if and only if there is a one-to-one correspondence between their vertices such that the corresponding angles are congruent and the lengths of corresponding sides are proportional.

Given the rectangles below, compare the area and perimeter.

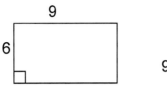

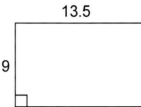

$A = LW$	$A = LW$	1. write formula
$A = (6)(9)$	$A = (9)(13.5)$	2. substitute known values
$A = 54$ sq. units	$A = 121.5$ sq. units	3. compute
$P = 2(L + W)$	$P = 2(L + W)$	1. write formula
$P = 2(6 + 9)$	$P = 2(9 + 13.5)$	2. substitute known values
$P = 30$ units	$P = 45$ units	3. compute

Notice that the areas relate to each other in the following manner:

Ratio of sides $9/13.5 = 2/3$

Multiply the first area by the square of the reciprocal $(3/2)^2$ to get the second area.

$$54 \times (3/2)^2 = 121.5$$

The perimeters relate to each other in the following manner:

Ratio of sides $9/13.5 = 2/3$

Multiply the perimeter of the first by the reciprocal of the ratio to get the perimeter of the second.

$$30 \times 3/2 = 45$$

Chords, Diameters, Radii, and Tangents

A tangent line intersects a circle in exactly one point. If a radius is drawn to that point, the radius will be perpendicular to the tangent.

A chord is a segment with endpoints on the circle. If a radius or diameter is perpendicular to a chord, the radius will cut the chord into two equal parts.

If two chords in the same circle have the same length, the two chords will have arcs that are the same length, and the two chords will be equidistant from the center of the circle.

Distance from the center to a chord is measured by finding the length of a segment from the center perpendicular to the chord.

Examples:

1.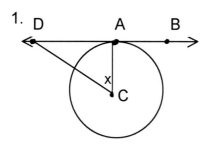

$\overrightarrow{DB}$ is tangent to $\odot C$ at A.
$m\angle ADC = 40°$. Find x.

$\overline{AC} \perp \overrightarrow{DB}$

A radius is $\perp$ to a tangent at the point of tangency.

$m\angle DAC = 90°$

Two segments that are $\perp$ form a $90°$ angle.

$40 + 90 + x = 180$

The sum of the angles of a triangle is $180°$.

$x = 50°$

Solve for x.

2.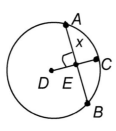

$\overline{CD}$ is a radius and
$\overline{CD} \perp$ chord $\overline{AB}$.
$\overline{AB} = 10$. Find x.

$x = \frac{1}{2}(10)$

$x = 5$

If a radius is $\perp$ to a chord, the radius bisects the chord.

Intersecting chords:

If two chords intersect inside a circle, each chord is divided into two smaller segments. The product of the lengths of the two segments formed from one chord equals the product of the lengths of the two segments formed from the other chord.

Intersecting tangent segments:

If two tangent segments intersect outside of a circle, the two segments have the same length.

Intersecting secant segments:

If two secant segments intersect outside a circle, a portion of each segment will lie inside the circle and a portion (called the exterior segment) will lie outside the circle.

The product of the length of one secant segment and the length of its exterior segment equals the product of the length of the other secant segment and the length of its exterior segment.

Tangent segments intersecting secant segments:

If a tangent segment and a secant segment intersect outside a circle, the square of the length of the tangent segment equals the product of the length of the secant segment and its exterior segment.

Examples:

1.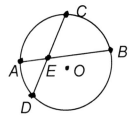

$\overline{AB}$ and $\overline{CD}$ are chords.
CE=10, ED=x, AE=5, EB=4

$(AE)(EB) = (CE)(ED)$ Since the chords intersect in the circle, the products of the segment pieces are equal.

$$5(4) = 10x$$
$$20 = 10x$$
$$x = 2$$ Solve for x.

2.

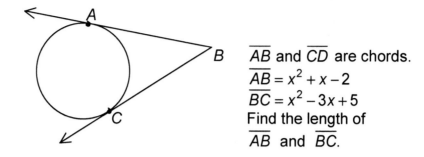

$\overline{AB}$ and $\overline{CD}$ are chords.
$\overline{AB} = x^2 + x - 2$
$\overline{BC} = x^2 - 3x + 5$
Find the length of
$\overline{AB}$ and $\overline{BC}$.

$\overline{AB} = x^2 + x - 2$

$\overline{BC} = x^2 - 3x + 5$

Given

$\overline{AB} = \overline{BC}$

Intersecting tangents are equal.

$x^2 + x - 2 = x^2 - 3x + 5$

Set the expressions equal to each other and solve.

$4x = 7$

$x = 1.75$

Substitute and solve.

$(1.75)^2 + 1.75 - 2 = \overline{AB}$ $\overline{AB} = \overline{BC} = 2.81$

Triangles

State and apply the relationships that exist when the altitude is drawn to the hypotenuse of a right triangle.

A **right triangle** is a triangle with one right angle. The side opposite the right angle is called the **hypotenuse**. The other two sides are the **legs**. An **altitude** is a line drawn from one vertex, perpendicular to the opposite side.

When an altitude is drawn to the hypotenuse of a right triangle, then the two triangles formed are similar to the original triangle and to each other.

Example:

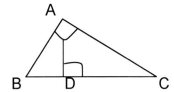

Given right triangle ABC with right angle at A, altitude AD drawn to hypotenuse BC at D.

$\triangle ABC \sim \triangle ABD \sim \triangle ACD$ The triangles formed are similar to each other and to the original right triangle.

Geometric Means

If a, b and c are positive numbers such that $\dfrac{a}{b} = \dfrac{b}{c}$ then b is called the **geometric mean** between a and c.

Example:

Find the geometric mean between 6 and 30.

$$\frac{6}{x} = \frac{x}{30}$$

$$x^2 = 180$$

$$x = \sqrt{180} = \sqrt{36 \cdot 5} = 6\sqrt{5}$$

The geometric mean is significant when the altitude is drawn to the hypotenuse of a right triangle.

The length of the altitude is the geometric mean between each segment of the hypotenuse,
    and
Each leg is the geometric mean between the hypotenuse and the segment of the hypotenuse that is adjacent to the leg.

Example:

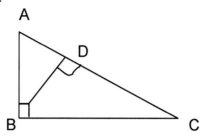

ΔABC is a right Δ
BD is the altitude of ΔABC
AB = 6
AC = 12
Find AD, CD, BD, and BC

$$\frac{12}{6} = \frac{6}{AD}$$

$$12(AD) = 36$$

$$AD = 3$$

$$\frac{3}{BD} = \frac{BD}{9}$$

$$(BD)^2 = 27$$

$$BD = \sqrt{27} = \sqrt{9 \cdot 3} = 3\sqrt{3}$$

$$\frac{12}{BC} = \frac{BC}{9}$$

$$(BC)^2 = 108$$

$$BC = \sqrt{108} = \sqrt{36 \cdot 3} = 6\sqrt{3}$$

$$CD = 12 - 3 = 9$$

Pythagorean Theorem

Pythagorean theorem states that the square of the length of the hypotenuse is equal to the sum of the squares of the lengths of the legs. Symbolically, this is stated as:

$$c^2 = a^2 + b^2$$

Given the right triangle below, find the missing side.

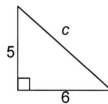

$$c^2 = a^2 + b^2$$ 1. write formula
$$c^2 = 5^2 + 6^2$$ 2. substitute known values
$$c^2 = 61$$ 3. take square root
$$c = \sqrt{61} \text{ or } 7.81$$ 4. solve

The converse of the Pythagorean Theorem states that if the square of one side of a triangle is equal to the sum of the squares of the other two sides, then the triangle is a right triangle.

Example:
Given $\triangle XYZ$, with sides measuring 12, 16 and 20 cm. Is this a right triangle?

$c^2 = a^2 + b^2$
20^2 ? $12^2 + 16^2$
400 ? $144 + 256$
$400 = 400$

Yes, the triangle is a right triangle.

This theorem can be expanded to determine if triangles are obtuse or acute.

If the square of the longest side of a triangle is greater than the sum of the squares of the other two sides, then the triangle is an obtuse triangle.
and
If the square of the longest side of a triangle is less than the sum of the squares of the other two sides, then the triangle is an acute triangle.

Example:
Given $\triangle LMN$ with sides measuring 7, 12, and 14 inches. Is the triangle right, acute, or obtuse?

14^2 ? $7^2 + 12^2$
196 ? $49 + 144$
$196 > 193$

Therefore, the triangle is obtuse.

30-60-90 and 45-45-90 Triangles

Given the special right triangles below, we can find the lengths of other special right triangles.

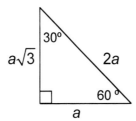 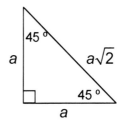

Sample problems:

1. 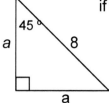 if $8 = a\sqrt{2}$ then $a = 8/\sqrt{2}$ or 5.657

2. 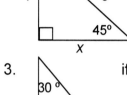 if $7 = a$ then $c = a\sqrt{2} = 7\sqrt{2}$ or 9.899

3. if $2a = 10$ then $a = 5$ and $x = a\sqrt{3} = 5\sqrt{3}$ or 8.66

Parallelograms

A **parallelogram** exhibits these properties.

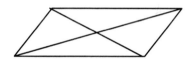

The diagonals bisect each other.
Each diagonal divides the parallelogram into two congruent triangles.
Both pairs of opposite sides are congruent.
Both pairs of opposite angles are congruent.
Two adjacent angles are supplementary.

Example 1:
Find the measures of the other three angles of a parallelogram if one angle measures 38°.

Since opposite angles are equal, there are two angles measuring 38°.

Since adjacent angles are supplementary, 180 - 38 = 142
so the other two angles measure 142 ° each.

$$
\begin{array}{r}
38 \\
38 \\
142 \\
+\ \underline{142} \\
360
\end{array}
$$

Example 2:
The measures of two adjacent angles of a parallelogram are $3x + 40$ and
$x + 70$.

Find the measures of each angle.

$$
\begin{aligned}
2(3x + 40) + 2(x + 70) &= 360 \\
6x + 80 + 2x + 140 &= 360 \\
8x + 220 &= 360 \\
8x &= 140 \\
x &= 17.5 \\
3x + 40 &= 92.5 \\
x + 70 &= 87.5
\end{aligned}
$$

Thus the angles measure $92.5°$, $92.5°$, $87.5°$, and $87.5°$.

Since a **rectangle** is a special type of parallelogram, it exhibits all the properties of a parallelogram. All the angles of a rectangle are right angles because of congruent opposite angles. Additionally, the diagonals of a rectangle are congruent.

A **rhombus** also has all the properties of a parallelogram. Additionally, its diagonals are perpendicular to each other and they bisect its angles.

A **square** has all the properties of a rectangle and a rhombus.

Example 1:

	True or false?
All squares are rhombuses.	True
All parallelograms are rectangles.	False - <u>some</u> parallelograms are rectangles
All rectangles are parallelograms.	True
Some rhombuses are squares.	True
Some rectangles are trapezoids.	False - only <u>one</u> pair of parallel sides
All quadrilaterals are parallelograms.	False -some quadrilaterals are parallelograms
Some squares are rectangles.	False - all squares are rectangles
Some parallelograms are rhombuses.	True

Example 2:

In rhombus $ABCD$ side $AB = 3x - 7$ and side $CD = x + 15$. Find the length of each side.

Since all the sides are the same length, $3x - 7 = x + 15$
$$2x = 22$$
$$x = 11$$

Since $3(11) - 7 = 25$ and $11 + 15 = 25$, each side measures 26 units.

Triangle Congruency

Use the SAS, ASA, and SSS postulates to show pairs of triangles congruent.

Two triangles can be proven congruent by comparing pairs of appropriate congruent corresponding parts.

SSS POSTULATE

If three sides of one triangle are congruent to three sides of another triangle, then the two triangles are congruent.

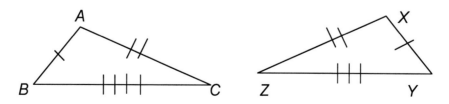

Since $AB \cong XY$, $BC \cong YZ$ and $AC \cong XZ$, then $\triangle ABC \cong \triangle XYZ$.

Example: Given isosceles triangle *ABC* with *D* the midpoint of base *AC*, prove the two triangles formed by *AD* are congruent.

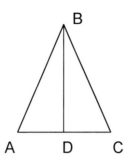

Proof:
1. Isosceles triangle *ABC*,
   D midpoint of base *AC*   Given
2. $AB \cong BC$ An isosceles △ has two congruent sides
3. $AD \cong DC$ Midpoint divides a line into two equal parts
4. $BD \cong BD$ Reflexive
5. △ *ABD* $\cong$ △*BCD*   SSS

SAS POSTULATE

If two sides and the included angle of one triangle are congruent to two sides and the included angle of another triangle, then the two triangles are congruent.

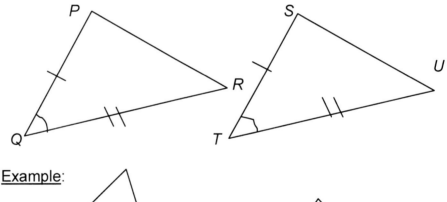

Example:

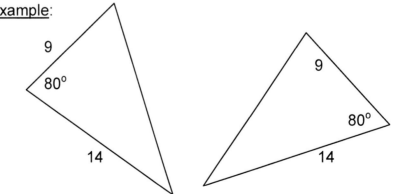

The two triangles are congruent by SAS.

ASA POSTULATE

If two angles and the included side of one triangle are congruent to two angles and the included side of another triangle, the triangles are congruent.

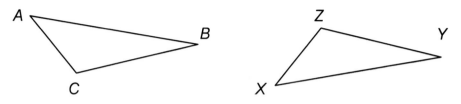

$\angle A \cong \angle X$, $\angle B \cong \angle Y$, $AB \cong XY$ then $\triangle ABC \cong \triangle XYZ$ by ASA

<u>Example 1</u>: Given two right triangles with one leg of each measuring 6 cm and the adjacent angle 37°, prove the triangles are congruent.

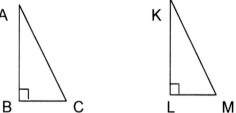

1. Right triangles *ABC* and *KLM*   Given
   $AB = KL = 6$ cm
   $\angle A = \angle K = 37°$

2. $AB \cong KL$ Figures with the
                                                        same measure
   $\angle A \cong \angle K$ are congruent

3. $\angle B \cong \angle L$ All right angles are
                                                        congruent.

4. $\triangle ABC \cong \triangle KLM$ ASA

<u>Example 2</u>:
What method would you use to prove the triangles congruent?

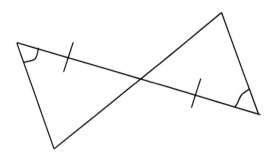

ASA because vertical angles are congruent.

AAS THEOREM

If two angles and a non-included side of one triangle are congruent to the corresponding parts of another triangle, then the triangles are congruent.

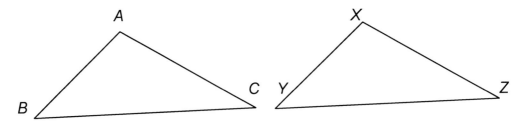

$\angle B \cong \angle Y$, $\angle C \cong \angle Z$, $AC \cong XZ$, then $\triangle ABC \cong \triangle XYZ$ by AAS.

We can derive this theorem because if two angles of the triangles are congruent, then the third angle must also be congruent. Therefore, we can uses the ASA postulate.

HL THEOREM

If the hypotenuse and a leg of one right triangle are congruent to the corresponding parts of another right triangle, the triangles are congruent.

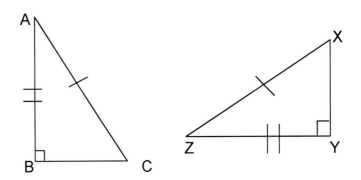

Since $\angle B$ and $\angle Y$ are right angles and $AC \cong XZ$ (hypotenuse of each triangle), $AB \cong YZ$ (corresponding leg of each triangle), then $\triangle ABC \cong \triangle XYZ$ by HL.

<u>Example</u>: What method would you use to prove the triangles congruent?

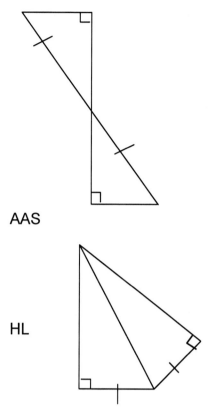

AAS

HL

Two triangles are congruent if each of the three angles and three sides of one triangle match up in a one-to-one fashion with congruent angles and sides of the second triangle. In order to see how the sides and angles match up, it is sometimes necessary to imagine rotating or reflecting one of the triangles so the two figures are oriented in the same position.

There are shortcuts to the above procedure for proving two triangles congruent.

Side-Side-Side (SSS) Congruence--If the three sides of one triangle match up in a one-to-one congruent fashion with the three sides of the other triangle, then the two triangles are congruent. With SSS it is not necessary to even compare the angles; they will automatically be congruent.

Angle-Side-Angle (ASA) Congruence--If two angles of one triangle match up in a one-to-one congruent fashion with two angles in the other triangle and if the sides between the two angles are also congruent, then the two triangles are congruent. With ASA the sides that are used for congruence must be located between the two angles used in the first part of the proof.

Side-Angle-Side (SAS) Congruence--If two sides of one triangle match up in a one-to-one congruent fashion with two sides in the other triangle and if the angles between the two sides are also congruent, then the two triangles are congruent. With SAS the angles that are used for congruence must be located between the two sides used in the first part of the proof.

In addition to SSS, ASA, and SAS, **Angle-Angle-Side (AAS)** is also a congruence shortcut.

AAS states that if two angles of one triangle match up in a one-to-one congruent fashion with two angle in the other triangle and if two sides that are not between the aforementioned sets of angles are also congruent, then the triangles are congruent. ASA and AAS are very similar; the only difference is where the congruent sides are located. If the sides are between the congruent sets of angles, use ASA. If the sides are not located between the congruent sets of angles, use AAS.

Hypotenuse-Leg (HL) is a congruence shortcut which can only be used with right triangles. If the hypotenuse and leg of one right triangle are congruent to the hypotenuse and leg of the other right triangle, then the two triangles are congruent.

Two triangles are overlapping if a portion of the interior region of one triangle is shared in common with all or a part of the interior region of the second triangle.

The most effective method for proving two overlapping triangles congruent is to draw the two triangles separated. Separate the two triangles and label all of the vertices using the labels from the original overlapping figures. Once the separation is complete, apply one of the congruence shortcuts: SSS, ASA, SAS, AAS, or HL.

Quadrilaterals

A parallelogram is a quadrilateral (four-sided figure) in which opposite sides are parallel. There are three shortcuts for proving that a quadrilateral is a parallelogram without directly showing that the opposite sides are parallel.

If the diagonals of a quadrilateral bisect each other, then the quadrilateral is also a parallelogram. Note that this shortcut only requires the diagonals to bisect each other; the diagonals do not need to be congruent.

If both pairs of opposite sides are congruent, then the quadrilateral is a parallelogram.

If both pairs of opposite angles are congruent, then the quadrilateral is a parallelogram.

If one pair of opposite sides are both parallel and congruent, then the quadrilateral is a parallelogram.

The following table illustrates the properties of each quadrilateral.

	Parallel Opposite Sides	Bisecting Diagonals	Equal Opposite Sides	Equal Opposite Angles	Equal Diagonals	All Sides Equal	All Angles Equal	Perpendicular Diagonals
Parallelogram	X	X	X	X				
Rectangle	X	X	X	X	X		X	
Rhombus	X	X	X	X		X		X
Square	X	X	X	X	X	X	X	X

Trapezoid Medians

A trapezoid is a quadrilateral with exactly one pair of parallel sides. A trapezoid is different from a parallelogram because a parallelogram has two pairs of parallel sides.

The two parallel sides of a trapezoid are called the bases, and the two non-parallel sides are called the legs. If the two legs are the same length, then the trapezoid is called isosceles.
The segment connecting the two midpoints of the legs is called the median. The median has the following two properties.

The median is parallel to the two bases.

The length of the median is equal to one-half the sum of the length of the two bases.

The segment joining the midpoints of two sides of a triangle is called a median. All triangles have three medians. Each median has the following two properties.

A median is parallel to the third side of the triangle.
The length of a median is one-half the length of the third side of the triangle.

Angle Bisectors

Every angle has exactly one ray which bisects the angle. If a point on such a bisector is located, then the point is equidistant from the two sides of the angle. Distance from a point to a side is measured along a segment which is perpendicular to the angle's side. The converse is also true. If a point is equidistant from the sides of an angle, then the point is on the bisector of the angle.

Every segment has exactly one line which is both perpendicular to and bisects the segment. If a point on such a perpendicular bisector is located, then the point is equidistant to the endpoints of the segment. The converse is also true. If a point is equidistant from the endpoints of a segments, then that point is on the perpendicular bisector of the segment.

If the three segments which bisect the three angles of a triangle are drawn, the segments will all intersect in a single point. This point is equidistant from all three sides of the triangle. Recall that the distance from a point to a side is measured along the perpendicular from the point to the side.

Triangle Medians

A median is a segment that connects a vertex to the midpoint of the side opposite from that vertex. Every triangle has exactly three medians.

An altitude is a segment which extends from one vertex and is perpendicular to the side opposite that vertex. In some cases, the side opposite from the vertex used will need to be extended in order for the altitude to form a perpendicular to the opposite side. The length of the altitude is used when referring to the height of the triangle.

Parallel Planes

If two planes are parallel and a third plane intersects the first two, then the three planes will intersect in two lines which are also parallel.

Given a line and a point which is not on the line but is in the same plane, then there is exactly one line through the point which is parallel to the given line and exactly one line through the point which is perpendicular to the given line.

Concurrency

If three or more segments intersect in a single point, the point is called a point of concurrency.

The following sets of special segments all intersect in points of concurrency.

1. Angle Bisectors
2. Medians
3. Altitudes
4. Perpendicular Bisectors

The points of concurrency can lie inside the triangle, outside the triangle, or on one of the sides of the triangle. The following table summarizes this information.

Possible Location(s) of the
Points of Concurrency

	Inside the Triangle	Outside the Triangle	On the Triangle
Angle Bisectors	x		
Medians	x		
Altitudes	x	x	x
Perpendicular Bisectors	x	x	x

Circles Inscribed in Triangles

A circle is inscribed in a triangle if the three sides of the triangle are each tangent to the circle. The center of an inscribed circle is called the incenter of the triangle. To find the incenter, draw the three angle bisectors of the triangle. The point of concurrency of the angle bisectors is the incenter or center of the inscribed circle. Each triangle has only one inscribed circle.

A circle is circumscribed about a triangle if the three vertices of the triangle are all located on the circle. The center of a circumscribed circle is called the circumcenter of the triangle. To find the circumcenter, draw the three perpendicular bisectors of the sides of the triangle. The point of concurrency of the perpendicular bisectors is the circumcenter or the center of the circumscribing circle. Each triangle has only one circumscribing circle.

Concurrency Points of Medians

A median is a segment which connects a vertex to the midpoint of the side opposite that vertex. Every triangle has three medians. The point of concurrency of the three medians is called the centroid.

The centroid divides each median into two segments whose lengths are always in the ratio of 1:2. The distance from the vertex to the centroid is always twice the distance from the centroid to the midpoint of the side opposite the vertex.

Circles – Relationships

If two circles have radii which are in a ratio of $a:b$, then the following ratios are also true for the circles.

The diameters are also in the ratio of $a:b$.
The circumferences are also in the ratio $a:b$.
The areas are in the ratio $a^2:b^2$, or the ratio of the areas is the square of the ratios of the radii.

COMPETENCY 0013   UNDERSTAND THE PRINCIPLES AND PROPERTIES OF COORDINATE GEOMETRY.

Distance Formula

The key to applying the distance formula is to understand the problem before beginning.

$$D = \sqrt{(x_2 - x_1)^2 + (y_2 - y_1)^2}$$

Sample Problem:

1. Find the perimeter of a figure with vertices at (4,5), ($^-$4,6) and ($^-$5, $^-$8).

The figure being described is a triangle. Therefore, the distance for all three sides must be found. Carefully, identify all three sides before beginning.

Side 1 = (4,5) to ($^-$4,6)

Side 2 = ($^-$4,6) to ($^-$5, $^-$8)

Side 3 = ($^-$5, $^-$8) to (4,5)

$$D_1 = \sqrt{(-4 - 4)^2 + (6 - 5)^2} = \sqrt{65}$$

$$D_2 = \sqrt{(^-5 - (^-4))^2 + (^-8 - 6)^2} = \sqrt{197}$$

$$D_3 = \sqrt{(4 - (^-5))^2 + (5 - (^-8)^2} = \sqrt{250} \text{ or } 5\sqrt{10}$$

$$\text{Perimeter} = \sqrt{65} + \sqrt{197} + 5\sqrt{10}$$

Distance- Point to Line

In order to accomplish the task of finding the distance from a given point to another given line the perpendicular line that intersects the point and line must be drawn and the equation of the other line written. From this information the point of intersection can be found. This point and the original point are used in the distance formula given below:

$$D = \sqrt{(x_2 - x_1)^2 + (y_2 - y_1)^2}$$

Sample Problem:

1. Given the point $(^-4,3)$ and the line $y = 4x + 2$, find the distance from the point to the line.

$y = 4x + 2$	1. Find the slope of the given line by solving for y.
$y = 4x + 2$	2. The slope is $4/1$, the perpendicular line will have a slope of $^-1/4$.
$y = \left(^-1/4\right)x + b$	3. Use the new slope and the given point to find the equation of the perpendicular line.
$3 = \left(^-1/4\right)\left(^-4\right) + b$	4. Substitute $(^-4,3)$ into the equation.
$3 = 1 + b$	5. Solve.
$2 = b$	6. Given the value for b, write the equation of the perpendicular line.
$y = \left(^-1/4\right)x + 2$	7. Write in standard form.
$x + 4y = 8$	8. Use both equations to solve by elimination to get the point of intersection.
$^-4x + y = 2$	
$\underline{x + 4y = 8}$	9. Multiply the bottom row by 4.
$^-4x + y = 2$	
$\underline{4x + 16y = 32}$	10. Solve.
$17y = 34$	
$y = 2$	
$y = 4x + 2$	11. Substitute to find the x value.
$2 = 4x + 2$	12. Solve.
$x = 0$	

(0,2) is the point of intersection. Use this point on the original line and the original point to calculate the distance between them.

$$D = \sqrt{(x_2 - x_1)^2 + (y_2 - y_1)^2}$$
where points are (0,2) and (-4,3).

$$D = \sqrt{(^-4 - 0)^2 + (3 - 2)^2}$$
1. Substitute.

$$D = \sqrt{(16) + (1)}$$
2. Simplify.

$$D = \sqrt{17}$$

Midpoint Formula

Midpoint Definition:

If a line segment has endpoints of (x_1, y_1) and (x_2, y_2), then the midpoint can be found using:

$$\left(\frac{x_1 + x_2}{2}, \frac{y_1 + y_2}{2} \right)$$

Sample problems:

1. Find the center of a circle with a diameter whose endpoints are (3, 7) and (−4, −5).

$$\text{Midpoint} = \left(\frac{3 + (-4)}{2}, \frac{7 + (-5)}{2} \right)$$

$$\text{Midpoint} = \left(\frac{-1}{2}, 1 \right)$$

2. Find the midpoint given the two points $\left(5, 8\sqrt{6} \right)$ and $\left(9, -4\sqrt{6} \right)$.

$$\text{Midpoint} = \left(\frac{5 + 9}{2}, \frac{8\sqrt{6} + (-4\sqrt{6})}{2} \right)$$

$$\text{Midpoint} = \left(7, 2\sqrt{6} \right)$$

Parallel Lines – Distance Between

The distance between two parallel lines, such as line *AB* and line *CD* as shown below is the line segment *RS*, the perpendicular between the two parallels.

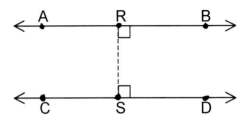

Sample Problem:

Given the geometric figure below, find the distance between the two parallel sides *AB* and *CD*.

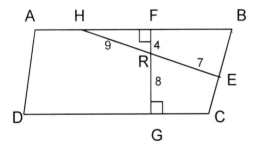

The distance *FG* is 12 units.

We refer to three-dimensional figures in geometry as **solids**. A solid is the union of all points on a simple closed surface and all points in its interior. A **polyhedron** is a simple closed surface formed from planar polygonal regions. Each polygonal region is called a **face** of the polyhedron. The vertices and edges of the polygonal regions are called the **vertices** and **edges** of the polyhedron.

We may form a cube from three congruent squares. However, if we tried to put four squares about a single vertex, their interior angle measures would add up to 360°; i.e., four edge-to-edge squares with a common vertex lie in a common plane and therefore cannot form a corner figure of a regular polyhedron.

There are five ways to form corner figures with congruent regular polygons:

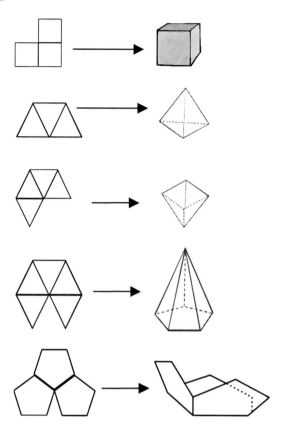

When creating a three-dimensional figure, if we know any two values of the vertices, faces, and edges, we can find the remaining value by using **Euler's Formula**: $V + F = E + 2$.

For example:

We want to create a pentagonal pyramid, and we know it has six vertices and six faces. Using Euler's Formula, we compute:

$$V + F = E + 2$$
$$6 + 6 = E + 2$$
$$12 = E + 2$$
$$10 = E$$

Thus, we know that our figure should have 10 edges.

Representing two- and three-dimensional geometric figures in the coordinate plane.

We can represent any two-dimensional geometric figure in the **Cartesian** or **rectangular coordinate system**. The Cartesian or rectangular coordinate system is formed by two perpendicular axes (coordinate axes): the X-axis and the Y-axis. If we know the dimensions of a two-dimensional, or planar, figure, we can use this coordinate system to visualize the shape of the figure.

Example: Represent an isosceles triangle with two sides of length 4.

Draw the two sides along the x- and y- axes and connect the points (vertices).

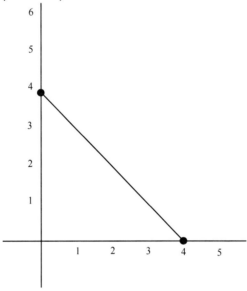

In order to represent three-dimensional figures, we need three coordinate axes (X, Y, and Z) which are all mutually perpendicular to each other. Since we cannot draw three mutually perpendicular axes on a two-dimensional surface, we use oblique representations.

Example: Represent a cube with sides of 2.

Once again, we draw three sides along the three axes to make things easier.

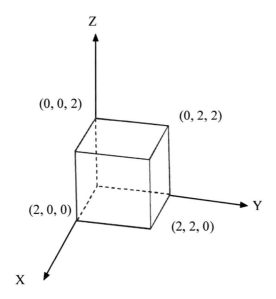

Each point has three coordinates (x, y, z).

Polar Plane

One way to graph points is in the rectangular coordinate system. In this system, the point (a,b) describes the point whose distance along the x-axis is "a" and whose distance along the y-axis is "b." The other method used to locate points is the **polar plane coordinate system**. This system consists of a fixed point called the pole or origin (labeled O) and a ray with O as the initial point called the polar axis. The ordered pair of a point P in the polar coordinate system is (r,θ), where $|r|$ is the distance from the pole and θ is the angle measure from the polar axis to the ray formed by the pole and point P. The coordinates of the pole are $(0,\theta)$, where θ is arbitrary. Angle θ can be measured in either degrees or in radians.

Sample problem:

1. Graph the point P with polar coordinates $(^-2, ^-45\ \text{degrees})$.

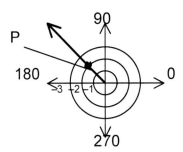

Draw $\theta = ^-45$ degrees in standard position. Since r is negative, locate the point $\left|^-2\right|$ units from the pole on the ray opposite the terminal side of the angle. Note that P can be represented by $(^-2, ^-45\ \text{degrees} + 180\ \text{degrees}) = (2, 135\ \text{degrees})$ or by $(^-2, ^-45\ \text{degrees} - 180\ \text{degrees}) = (2, ^-225\ \text{degrees})$.

2. Graph the point $P = \left(3, \dfrac{\pi}{4}\right)$ and show another graph that also represents the same point P.

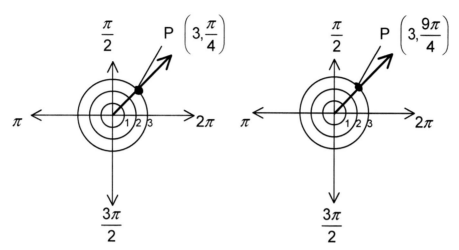

In the second graph, the angle 2π is added to $\dfrac{\pi}{4}$ to give the point $\left(3, \dfrac{9\pi}{4}\right)$.

It is possible that r be allowed to be negative. Now instead of measuring $|r|$ units along the terminal side of the angle, we would locate the point $\left|^-3\right|$ units from the pole on the ray opposite the terminal side. This would give the points $\left(^-3, \dfrac{5\pi}{4}\right)$ and $\left(^-3, \dfrac{^-3\pi}{4}\right)$.

COMPETENCY 0014 **UNDERSTAND THE PRINCIPLES AND PROPERTIES OF VECTOR AND TRANSFORMATIONAL GEOMETRIES.**

Occasionally, it is important to reverse the addition or subtraction process and express the single vector as the sum or difference of two other vectors. It may be critically important for a pilot to understand not only the air velocity but also the ground speed and the climbing speed.

Sample problem:

A pilot is traveling at an air speed of 300 mph and a direction of 20 degrees. Find the horizontal vector (ground speed) and the vertical vector (climbing speed).

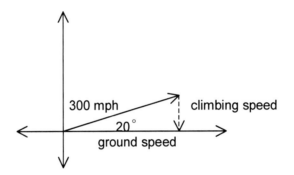

1. Draw sketch.
2. Use appropriate trigonometric ratio to calculate the component vectors.

To find the vertical vector:

$$\sin x = \frac{\text{opposite}}{\text{hypotenuse}}$$

$$\sin(20) = \frac{c}{300}$$

$$c = (.3420)(300)$$

$$c = 102.606$$

To find the horizontal vector:

$$\cos x = \frac{\text{adjacent}}{\text{hypotenuse}}$$

$$\cos(20) = \frac{g}{300}$$

$$g = (.9397)(300)$$

$$g = 281.908$$

Addition and Subtraction of Vectors

Vectors are used to measure displacement of an object or force.

Addition of vectors:

$$(a,b)+(c,d)=(a+c,b+d)$$

Addition Properties of vectors:

$$a+b=b+a$$
$$a+(b+c)=(a+b)+c$$
$$a+0=a$$
$$a+(^-a)=0$$

Subtraction of vectors:

$$a-b=a+(^-b) \text{ therefore,}$$
$$a-b=(a_1,a_2)+(^-b_1,^-b_2) \text{ or}$$
$$a-b=(a_1-b_1,a_2-b_2)$$

Sample problem:

If $a=(4,^-1)$ and $b=(^-3,6)$, find $a+b$ and $a-b$.

Using the rule for addition of vectors:

$$(4,^-1)+(^-3,6)=(4+(^-3),^-1+6)$$
$$=(1,5)$$

Using the rule for subtraction of vectors:

$$(4,^-1)-(^-3,6)=(4-(^-3),^-1-6)$$
$$=(7,^-7)$$

Scalar Multiplication

The dot product $a \cdot b$:

$$a = (a_1, a_2) = a_1 i + a_2 j \quad \text{and} \quad b = (b_1, b_2) = b_1 i + b_2 j$$

$$a \cdot b = a_1 b_1 + a_2 b_2$$

$a \cdot b$ is read "a dot b". Dot products are also called scalar or inner products. When discussing dot products, it is important to remember that "a dot b" is not a vector, but a real number.

Properties of the dot product:
$$a \cdot a = |a|^2$$
$$a \cdot b = b \cdot a$$
$$a \cdot (b + c) = a \cdot b + a \cdot c$$
$$(ca) \cdot b = c(a \cdot b) = a \cdot (cb)$$
$$0 \cdot a = 0$$

Sample problems:

Find the dot product.

1. $a = (5,2), b = (^-3,6)$
    $a \cdot b = (5)(^-3) + (2)(6)$
    $\qquad = {}^-15 + 12$
    $\qquad = {}^-3$

2. $a = (5i + 3j), b = (4i - 5j)$
    $a \cdot b = (5)(4) + (3)(^-5)$
    $\qquad = 20 - 15$
    $\qquad = 5$

3. The magnitude and direction of a constant force are given by $a = 4i + 5j$. Find the amount of work done if the point of application of the force moves from the origin to the point $P(7,2)$.

The work W done by a constant force a as its point of application moves along a vector b is $W = a \cdot b$.

Sketch the constant force vector a and the vector b.

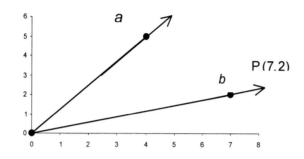

$b = (7,2) = 7i + 2j$

Use the definition of work done to solve.

$$W = a \cdot b$$
$$= (4i + 5j)(7i + 2j)$$
$$= (4)(7) + (5)(2)$$
$$= (28) + (10)$$
$$= 38$$

Problem Solving

Vectors are used often in navigation of ships and aircraft as well as in force and work problems.

Sample problem:

1. An airplane is flying with a heading of 60 degrees east of north at 450 mph. A wind is blowing at 37 mph from the north. Find the plane's ground speed to the nearest mph and direction to the nearest degree.

Draw a sketch.

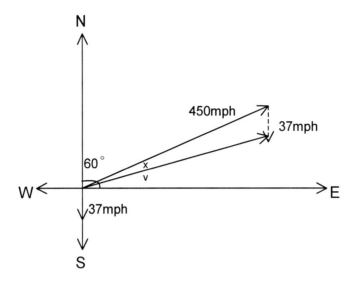

Use the law of cosines to find the ground speed.

$$a^2 = b^2 + c^2 - 2bc \cos A$$

$$|v|^2 = 450^2 + 37^2 - 2(450)(37)\cos 60$$

$$|v|^2 = 187,219$$

$$|v| = 432.688$$

$$|v| \approx 433 \text{ mph}$$

Use the law of sines to find the measure of x.

$$\frac{\sin A}{a} = \frac{\sin C}{c}$$

$$\frac{\sin 60}{433} = \frac{\sin x}{37}$$

$$\sin x = \frac{37(\sin 60)}{433}$$

$$x = 4.24$$

$$x \approx 4 \text{ degrees}$$

The plane's actual course is $60 + 4 = 64$ degrees east of north and the ground speed is approximately 433 miles per hour.

The plane's actual course is $60 + 4 = 64$ degrees east of north and the ground speed is approximately 433 miles per hour.

A **transformation** is a change in the position, shape, or size of a geometric figure. **Transformational geometry** is the study of manipulating objects by flipping, twisting, turning and scaling. **Symmetry** is exact similarity between two parts or halves, as if one were a mirror image of the other.

There are four basic transformational symmetries that can be used in tessellations: **translation, rotation, reflection,** and **glide reflection**. The transformation of an object is called its image. If the original object was labeled with letters, such as *ABCD*, the image may be labeled with the same letters followed by a prime symbol, *A′B′C′D′*.

A **translation** is a transformation that "slides" an object a fixed distance in a given direction. The original object and its translation have the same shape and size, and they face in the same direction.

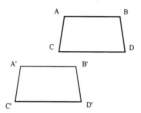

An example of a translation in architecture would be stadium seating. The seats are the same size and the same shape and face in the same direction.

A **rotation** is a transformation that turns a figure about a fixed point called the center of rotation.

An object and its rotation are the same shape and size, but the figures may be turned in different directions. Rotations can occur in either a clockwise or a counterclockwise direction.

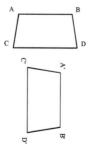

Rotations can be seen in wallpaper and art, and a Ferris wheel is an example of rotation.

An object and its **reflection** have the same shape and size, but the figures face in opposite directions.

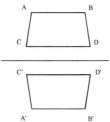

The line (where a mirror may be placed) is called the **line of reflection**. The distance from a point to the line of reflection is the same as the distance from the point's image to the line of reflection.

A **glide reflection** is a combination of a reflection and a translation.

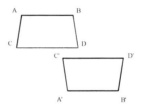

The tessellation below is a combination of the four types of transformational symmetry we have discussed:

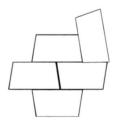

Plot the given ordered pairs on a coordinate plane and join them in the given order, then join the first and last points.

(-3, -2), (3, -2), (5, -4), (5, -6), (2, -4), (-2, -4), (-5, -6), (-5, -4)

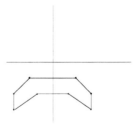

Increase all y-coordinates by 6.

(-3, 4), (3, 4), (5, 2), (5, 0), (2, 2), (-2, 2), (-5, 0), (-5, 2)

Plot the points and join them to form a second figure.

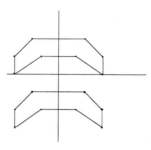

A figure on a coordinate plane can be translated by changing the ordered pairs.

SUBAREA IV.   TRIGONOMETRY AND CALCULUS

COMPETENCY 0015   UNDERSTAND THE PRINCIPLES AND
                         PROPERTIES OF AND RELATIONSHIPS
                         INVOLVING TRIGONOMETRIC FUNCTIONS AND
                         THEIR GRAPHIC REPRESENTATIONS.

The unit circle is a circle with a radius of one centered at (0,0) on
the coordinate plane. Thus, any ray from the origin to a point on
the circle forms an angle, t, with the positive x-axis. In addition, the
ray from the origin to a point on the circle in the first quadrant forms
a right triangle with a hypotenuse measuring one unit. Applying the
Pythagorean theorem, $x^2 + y^2 = 1$. Because the reflections of the
triangle about both the x- and y-axis are on the unit circle and $x^2 =
(-x)^2$ for all x values, the formula holds for all points on the unit
circle.

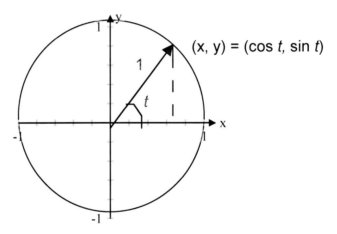

Note that because the length of the hypotenuse of the right
triangles formed in the unit circle is one, the point on the unit circle
that forms the triangle is (cos t, sin t). In other words, for any point
on the unit circle, the x value represents the cosine of t, the y value
represents the sine of t, and the ratio of the y-coordinate to the x-
coordinate is the tangent of t.

The unit circle illustrates several properties of trigonometric
functions. For example, applying the Pythagorean theorem to the
unit circle yields the equation $\cos^2(t) + \sin^2(t) = 1$. In addition, the
unit circle reveals the periodic nature of trigonometric functions.
When we increase the angle t beyond 2π radians or 360 degrees,
the values of x and y coordinates on the unit circle remain the
same. Thus, the sine and cosine values repeat with each
revolution. The unit circle also reveals the range of the sine and
cosine functions. The values of sine and cosine are always
between one and negative one.

Finally, the unit circle shows that when the x coordinate of a point on the circle is zero, the tangent function is undefined. Thus, tangent is undefined at the angles $\frac{\pi}{2}, \frac{3\pi}{2}$ and the corresponding angles in all subsequent revolutions.

Given triangle right ABC, the adjacent side and opposite side can be identified for each angle A and B.

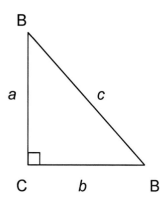

Looking at angle A, it can be determined that side *b* is adjacent to angle A and side *a* is opposite angle A.

If we now look at angle B, we see that side *a* is adjacent to angle *b* and side *b* is opposite angle B.

The longest side (opposite the 90 degree angle) is always called the hypotenuse.

The basic trigonometric ratios are listed below:

Sine = opposite / hypotenuse   Cosine = adjacent / hypotenuse   Tangent = opposite / adjacent

Sample problem:

1. Use triangle ABC to find the sin, cos and tan for angle A.

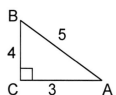

sin A= 4/5
cos A= 3/5
tan A= 4/3

Trigonometric Functions

It is easiest to graph trigonometric functions when using a calculator by making a table of values.

DEGREES

	0	30	45	60	90	120	135	150	180	210	225	240	270	300	315	330	360
sin	0	.5	.71	.87	1	.87	.71	.5	0	-.5	-.71	-.87	-1	-.87	-.71	-.5	0
cos	1	.87	.71	.5	0	-.5	-.71	-.87	-1	-.87	-.71	-.5	0	.5	.71	.87	1
tan	0	.58	1	1.7	--	-1.7	-1	-.58	0	.58	1	1.7	--	-1.7	-1	-.58	0

| 0 | $\dfrac{\pi}{6}$ | $\dfrac{\pi}{4}$ | $\dfrac{\pi}{3}$ | $\dfrac{\pi}{2}$ | $\dfrac{2\pi}{3}$ | $\dfrac{3\pi}{4}$ | $\dfrac{5\pi}{6}$ | π | $\dfrac{7\pi}{6}$ | $\dfrac{5\pi}{4}$ | $\dfrac{4\pi}{3}$ | $\dfrac{3\pi}{2}$ | $\dfrac{5\pi}{3}$ | $\dfrac{7\pi}{4}$ | $\dfrac{11\pi}{6}$ | 2π |

RADIANS

Remember the graph always ranges from +1 to ⁻1 for sine and cosine functions unless noted as the coefficient of the function in the equation. For example, $y = 3\cos x$ has an amplitude of 3 units from the center line (0). Its maximum and minimum points would be at +3 and ⁻3.

Tangent is not defined at the values 90 and 270 degrees or $\dfrac{\pi}{2}$ and $\dfrac{3\pi}{2}$. Therefore, vertical asymptotes are drawn at those values.

The inverse functions can be graphed in the same manner using a calculator to create a table of values.

Transformations of trigonometric functions change aspects of the graph - amplitude, period, frequency, or range - but do not change the basic nature of the curve. The two basic types of transformations are output based and input based. Output based transformations affect the amplitude and range of the graph of a function, but do not change the period. Input based transformations, on the other hand, do not change the range, but do affect the period.

Output based transformations include vertical shifts, rotations, and stretches. The following are examples of output based transformations of the function $f(x) = \sin x$ with the corresponding graphs.

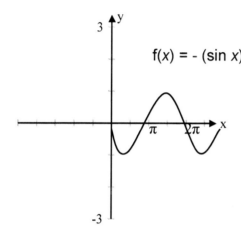

$f(x) = - (\sin x)$

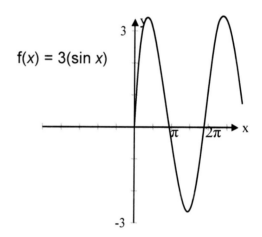

$f(x) = 3(\sin x)$

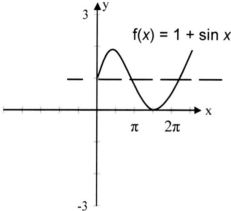

$f(x) = 1 + \sin x$

Note the graph of $f(x) = - (\sin x)$ is a reflection of $f(x) = \sin x$ about the x-axis. The graph of $f(x) = 3 (\sin x)$ is a vertical stretch of $f(x) = \sin x$ by a factor of 3 that produces an increase in amplitude or range. The graph of $f(x) = 1 + \sin x$ is a 1 unit vertical shift of $f(x) = \sin x$. In all cases, the period of the graph is 2π, the same as the period of $f(x) = \sin x$.

Input based transformations include expansions, compressions, and horizontal shifts. The following are examples of input based transformations of the function f(x) = sin x with the corresponding graphs.

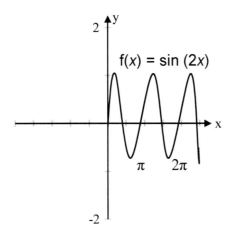

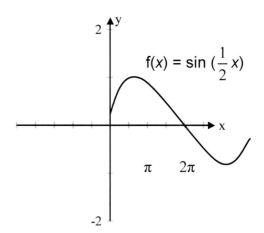

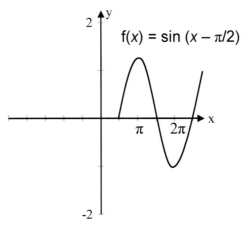

Note the graph of f(x) = sin (2x) is a compression of f(x) = sin x by a factor of 2 where the period is π rather that 2π. The graph of f(x) = sin (½x) is an expansion by a factor of 2 where the period is 4π rather that 2π. The graph of f(x) = sin (x − π/2) is a π/2 unit horizontal shift of f(x) = sin x and the period is unchanged. In all cases, the range of the graph is 1, the same as the range of f(x) = sin x.

Given 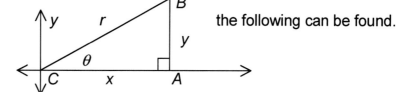 the following can be found.

Trigonometric Functions:

$$\sin\theta = \frac{y}{r} \qquad\qquad \csc\theta = \frac{r}{y}$$

$$\cos\theta = \frac{x}{r} \qquad\qquad \sec\theta = \frac{r}{x}$$

$$\tan\theta = \frac{y}{x} \qquad\qquad \cot\theta = \frac{x}{y}$$

Sample problem:

1. Prove that $\sec\theta = \dfrac{1}{\cos\theta}$.

$\sec\theta = \dfrac{1}{\dfrac{x}{r}}$ Substitution definition of cosine.

$\sec\theta = \dfrac{1 \times r}{\dfrac{x}{r} \times r}$ Multiply by $\dfrac{r}{r}$.

$\sec\theta = \dfrac{r}{x}$ Substitution.

$\sec\theta = \sec\theta$ Substitute definition of $\dfrac{r}{x}$.

$\sec\theta = \dfrac{1}{\cos\theta}$ Substitute.

2. Prove that $\sin^2 + \cos^2 = 1$.

$\left(\dfrac{y}{r}\right)^2 + \left(\dfrac{x}{r}\right)^2 = 1$ Substitute definitions of sin and cos.

$\dfrac{y^2 + x^2}{r^2} = 1$ $x^2 + y^2 = r^2$ Pythagorean formula.

$\dfrac{r^2}{r^2} = 1$ Simplify.

$1 = 1$ Substitute.

$\sin^2\theta + \cos^2\theta = 1$

Practice problems: Prove each identity.

1. $\cot\theta = \dfrac{\cos\theta}{\sin\theta}$ 2. $1 + \cot^2\theta + \csc^2\theta$

There are two methods that may be used to prove trigonometric identities. One method is to choose one side of the equation and manipulate it until it equals the other side. The other method is to replace expressions on both sides of the equation with equivalent expressions until both sides are equal.

The Reciprocal Identities

$\sin x = \dfrac{1}{\csc x}$ $\sin x \csc x = 1$ $\csc x = \dfrac{1}{\sin x}$

$\cos x = \dfrac{1}{\sec x}$ $\cos x \sec x = 1$ $\sec x = \dfrac{1}{\cos x}$

$\tan x = \dfrac{1}{\cot x}$ $\tan x \cot x = 1$ $\cot x = \dfrac{1}{\tan x}$

$\tan x = \dfrac{\sin x}{\cos x}$ $\cot x = \dfrac{\cos x}{\sin x}$

The Pythagorean Identities

$$\sin^2 x + \cos^2 x = 1 \qquad 1 + \tan^2 x = \sec^2 x \qquad 1 + \cot^2 x = \csc^2 x$$

Sample problems:

1. Prove that $\cot x + \tan x = (\csc x)(\sec x)$.

$\dfrac{\cos x}{\sin x} + \dfrac{\sin x}{\cos x}$ Reciprocal identities.

$\dfrac{\cos^2 x + \sin^2 x}{\sin x \cos x}$ Common denominator.

$\dfrac{1}{\sin x \cos x}$ Pythagorean identity.

$\dfrac{1}{\sin x} \times \dfrac{1}{\cos x}$

$\csc x(\sec x) = \csc x(\sec x)$ Reciprocal identity, therefore,
$\cot x + \tan x = \csc x(\sec x)$

2. Prove that $\dfrac{\cos^2 \theta}{1 + 2\sin\theta + \sin^2 \theta} = \dfrac{\sec\theta - \tan\theta}{\sec\theta + \tan\theta}$.

$\dfrac{1 - \sin^2 \theta}{(1 + \sin\theta)(1 + \sin\theta)} = \dfrac{\sec\theta - \tan\theta}{\sec\theta + \tan\theta}$ Pythagorean identity factor denominator.

$\dfrac{1 - \sin^2 \theta}{(1 + \sin\theta)(1 + \sin\theta)} = \dfrac{\dfrac{1}{\cos\theta} - \dfrac{\sin\theta}{\cos\theta}}{\dfrac{1}{\cos\theta} + \dfrac{\sin\theta}{\cos\theta}}$ Reciprocal identities.

$\dfrac{(1 - \sin\theta)(1 + \sin\theta)}{(1 + \sin\theta)(1 + \sin\theta)} = \dfrac{\dfrac{1 - \sin\theta}{\cos\theta}(\cos\theta)}{\dfrac{1 + \sin\theta}{\cos\theta}(\cos\theta)}$ Factor $1 - \sin^2 \theta$.

Multiply by $\dfrac{\cos\theta}{\cos\theta}$.

$\dfrac{1 - \sin\theta}{1 + \sin\theta} = \dfrac{1 - \sin\theta}{1 + \sin\theta}$ Simplify.

$\dfrac{\cos^2 \theta}{1 + 2\sin\theta + \sin^2 \theta} = \dfrac{\sec\theta - \tan\theta}{\sec\theta + \tan\theta}$

COMPETENCY 0016   **UNDERSTAND AND APPLY THE PRINCIPLES AND TECHNIQUES OF TRIGONOMETRY TO MODEL AND SOLVE PROBLEMS.**

In order to solve a right triangle using trigonometric functions it is helpful to identify the given parts and label them. Usually more than one trigonometric function may be appropriately applied.

Some items to know about right triangles:

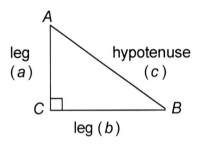

Given angle A, the side labeled leg (a) Is adjacent angle A. And the side labeled leg (b) is opposite to angle A.

Sample problem:

1. Find the missing side.

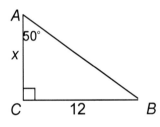

1. Identify the known values. Angle $A = 50$ degrees and the side opposite the given angle is 12. The missing side is the adjacent leg.

$$\tan A = \frac{\text{opposite}}{\text{adjacent}}$$

2. The information suggests the use of the tangent function

$$\tan 50 = \frac{12}{x}$$

3. Write the function.

$$1.192 = \frac{12}{x}$$

4. Substitute.

$$x(1.192) = 12$$

5. Solve.

$$x = 10.069$$

Remember that since angle A and angle B are complimentary, then angle $B = 90 - 50$ or 40 degrees.

Using this information we could have solved for the same side only this time it is the leg opposite from angle B.

$\tan B = \dfrac{\text{opposite}}{\text{adjacent}}$ 1. Write the formula.

$\tan 40 = \dfrac{x}{12}$ 2. Substitute.

$12(.839) = x$ 3. Solve.

$10.069 \approx x$

Now that the two sides of the triangle are known, the third side can be found using the Pythagorean Theorem.

Use the basic trigonometric ratios of sine, cosine and tangent to solve for the missing sides of right triangles when given at least one of the acute angles.

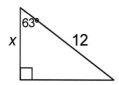   In the triangle ABC, an acute angle of 63 degrees and the length of the hypotenuse (12). The missing side is the one adjacent to the given angle.

The appropriate trigonometric ratio to use would be cosine since we are looking for the adjacent side and we have the length of the hypotenuse.

$\text{Cos}x = \dfrac{\text{adjacent}}{\text{hypotenuse}}$ 1. Write formula.

$\text{Cos } 63 = \dfrac{x}{12}$ 2. Substitute known values.

$0.454 = \dfrac{x}{12}$

$x = 5.448$ 3. Solve.

Law of Cosines

Definition: For any triangle ABC, when given two sides and the included angle, the other side can be found using one of the formulas below:

$$a^2 = b^2 + c^2 - (2bc)\cos A$$
$$b^2 = a^2 + c^2 - (2ac)\cos B$$
$$c^2 = a^2 + b^2 - (2ab)\cos C$$

Similarly, when given three sides of a triangle, the included angles can be found using the derivation:

$$\cos A = \frac{b^2 + c^2 - a^2}{2bc}$$

$$\cos B = \frac{a^2 + c^2 - b^2}{2ac}$$

$$\cos C = \frac{a^2 + b^2 - c^2}{2ab}$$

Sample problem:

1. Solve triangle ABC, if angle $B = 87.5°$, $a = 12.3$, and $c = 23.2$. (Compute to the nearest tenth).

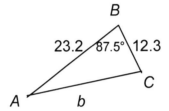

1. Draw and label a sketch.

Find side b.

$$b^2 = a^2 + c^2 - (2ac)\cos B$$

2. Write the formula.

$$b^2 = (12.3)^2 + (23.2)^2 - 2(12.3)(23.2)(\cos 87.5)$$ 3. Substitute.

$$b^2 = 664.636$$

$$b = 25.8 \ \text{(rounded)}$$

4. Solve.

Use the law of sines to find angle A.

$$\frac{\sin A}{a} = \frac{\sin B}{b}$$

1. Write formula.

$$\frac{\sin A}{12.3} = \frac{\sin 87.5}{25.8} = \frac{12.29}{25.8}$$

2. Substitute.

$$\sin A = 0.47629$$

3. Solve.

Angle $A = 28.4$

Therefore, angle $C = 180 - (87.5 + 28.4)$
$$= 64.1$$

2. Solve triangle ABC if $a = 15$, $b = 21$, and $c = 18$. (Round to the nearest tenth).

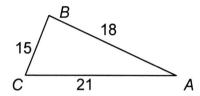

1. Draw and label a sketch.

Find angle A.

$$\cos A = \frac{b^2 + c^2 - a^2}{2bc}$$

2. Write formula.

$$\cos A = \frac{21^2 + 18^2 - 15^2}{2(21)(18)}$$

3. Substitute.

$\cos A = 0.714$

4. Solve.

Angle $A = 44.4$

Find angle B.

$$\cos B = \frac{a^2 + c^2 - b^2}{2ac}$$

5. Write formula.

$$\cos B = \frac{15^2 + 18^2 - 21^2}{2(15)(18)}$$

6. Substitute.

$\cos B = 0.2$

7. Solve.

Angle $B = 78.5$

Therefore, angle $C = 180 - (44.4 + 78.5)$
$$= 57.1$$

Law of Sines

Definition: For any triangle ABC, where a, b, and c are the lengths of the sides opposite angles A, B, and C respectively.

$$\frac{\sin A}{a} = \frac{\sin B}{b} = \frac{\sin C}{c}$$

Sample problem:

1. An inlet is 140 feet wide. The lines of sight from each bank to an approaching ship are 79 degrees and 58 degrees. What are the distances from each bank to the ship?

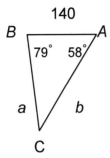

1. Draw and label a sketch.

2. The missing angle is
   $180 - (79 + 58) = 43$

   □ C=43 degrees.

$$\frac{\sin A}{a} = \frac{\sin B}{b} = \frac{\sin C}{c}$$

3. Write formula.

Side opposite 79 degree angle:

$$\frac{\sin 79}{b} = \frac{\sin 43}{140}$$

4. Substitute.

$$b = \frac{140(.9816)}{.6820}$$

5. Solve.

$b \approx 201.501$ feet

Side opposite 58 degree angle:

$$\frac{\sin 58}{a} = \frac{\sin 43}{140}$$

6. Substitute.

$$a = \frac{140(.848)}{.6820}$$

7. Solve.

$a \approx 174.076$ feet

Unlike trigonometric identities that are true for all values of the defined variable, trigonometric equations are true for some, but not all, of the values of the variable. Most often trigonometric equations are solved for values between 0 and 360 degrees or 0 and 2π radians.

Some algebraic operation, such as squaring both sides of an equation, will give you extraneous answers. You must remember to check all solutions to be sure that they work.

Sample problems:

1. Solve: $\cos x = 1 - \sin x$ if $0 \le x < 360$ degrees.

   $\cos^2 x = (1 - \sin x)^2$ 1. square both sides

   $1 - \sin^2 x = 1 - 2\sin x + \sin^2 x$ 2. substitute

   $0 = {}^-2\sin x + 2\sin^2 x$ 3. set $=$ to 0

   $0 = 2\sin x({}^-1 + \sin x)$ 4. factor

   $2\sin x = 0$ ${}^-1 + \sin x = 0$ 5. set each factor $= 0$

   $\sin x = 0$ $\sin x = 1$ 6. solve for $\sin x$

   $x = 0$ or 180 $x = 90$ 7. find value of at x

The solutions appear to be 0, 90 and 180. Remember to check each solution and you will find that 180 does not give you a true equation. Therefore, the only solutions are 0 and 90 degrees.

2. Solve: $\cos^2 x = \sin^2 x$ if $0 \le x < 2\pi$

   $\cos^2 x = 1 - \cos^2 x$ 1. substitute

   $2\cos^2 x = 1$ 2. simplify

   $\cos^2 x = \dfrac{1}{2}$ 3. divide by 2

   $\sqrt{\cos^2 x} = \pm\sqrt{\dfrac{1}{2}}$ 4. take square root

   $\cos x = \dfrac{\pm\sqrt{2}}{2}$ 5. rationalize denominator

   $x = \dfrac{\pi}{4}, \dfrac{3\pi}{4}, \dfrac{5\pi}{4}, \dfrac{7\pi}{4}$

COMPETENCY 0017 **UNDERSTAND THE PRINCIPLES AND PROPERTIES OF LIMITS, CONTINUITY, AND AVERAGE RATES OF CHANGE.**

The limit of a function is the y value that the graph approaches as the x values approach a certain number. To find a limit there are two points to remember.

1. Factor the expression completely and cancel all common factors in fractions.
2. Substitute the number to which the variable is approaching. In most cases this produces the value of the limit.

If the variable in the limit is approaching ∞, factor and simplify first; then examine the result. If the result does not involve a fraction with the variable in the denominator, the limit is usually also equal to ∞. If the variable is in the denominator of the fraction, the denominator is getting larger which makes the entire fraction smaller. In other words the limit is zero.

Examples:

1. $\lim\limits_{x \to {}^-3} \dfrac{x^2 + 5x + 6}{x + 3} + 4x$ Factor the numerator.

   $\lim\limits_{x \to {}^-3} \dfrac{(x+3)(x+2)}{(x+3)} + 4x$ Cancel the common factors.

   $\lim\limits_{x \to {}^-3} (x+2) + 4x$ Substitute $^-3$ for x.

   $({}^-3 + 2) + 4({}^-3)$ Simplify.

   $^-1 + {}^-12$

   $^-13$

2. $\lim\limits_{x \to \infty} \dfrac{2x^2}{x^5}$ Cancel the common factors.

   $\lim\limits_{x \to \infty} \dfrac{2}{x^3}$

   Since the denominator is getting larger, the entire fraction is getting smaller. The fraction is getting close to zero.

   $\dfrac{2}{\infty^3}$

Practice problems:

1. $\lim\limits_{x \to \pi} 5x^2 + \sin x$

2. $\lim\limits_{x \to -4} = \dfrac{x^2 + 9x + 20}{x + 4}$

L'Hopital's Rule

After simplifying an expression to evaluate a limit, substitute the value that the variable approaches. If the substitution results in either $0/0$ or ∞/∞, use L'Hopital's rule to find the limit.

L'Hopital's rule states that you can find such limits by taking the derivative of the numerator and the derivative of the denominator, and then finding the limit of the resulting quotient.

Examples:

1. $\displaystyle\lim_{x \to \infty} \frac{3x - 1}{x^2 + 2x + 3}$ 　　　No factoring is possible.

$$\frac{3\infty - 1}{\infty^2 + 2\infty + 3}$$ 　　　Substitute ∞ for x.

$$\frac{\infty}{\infty}$$

Since a constant times infinity is still a large number, $3(\infty) = \infty$.

$$\lim_{x \to \infty} \frac{3}{2x + 2}$$

To find the limit, take the derivative of the numerator and denominator.

$$\frac{3}{2(\infty) + 2}$$ 　　　Substitute ∞ for x again.

$$\frac{3}{\infty}$$

$$0$$

Since the denominator is a very large number, the fraction is getting smaller. Thus the limit is zero.

2. $\lim\limits_{x \to 1} \dfrac{\ln x}{x-1}$

Substitute 1 for x.

$\dfrac{\ln 1}{1-1}$

The $\ln 1 = 0$

$\dfrac{0}{0}$

To find the limit, take the derivative of the numerator and denominator.

$\lim\limits_{x \to 1} \dfrac{\frac{1}{x}}{1}$

Substitute 1 for x again.

$\dfrac{\frac{1}{1}}{1}$

Simplify. The limit is one.

1

Practice problems:

1. $\lim\limits_{x \to \infty} \dfrac{x^2 - 3}{x}$

2. $\lim\limits_{x \to \frac{\pi}{2}} \dfrac{\cos x}{x - \frac{\pi}{2}}$

The concept of limit applies when examining the way a function behaves close to a particular point of interest. Evaluating the continuity of a function and determining the nature of discontinuous functions is one application of the concept of limit. We also use limits to determine the asymptotes of the graph of a function.

If the limit of a function as x approaches a equals the value of the function at a, then the function is continuous at a. In other words, a function is continuous at a number a if

$$\lim\limits_{x \to a} f(x) = f(a).$$

The function f is continuous at a if the function satisfies the following three criteria.

1. f(a) is defined
2. The limit of f(x) as x approaches a exists.
3. The limit of f(x) as x approaches a equals f(a).

Example 1: $f(x) = \dfrac{x}{x-2}$

This function is discontinuous at x = 2 because f(2) is not defined.

Example 2: $f(x) = \dfrac{x^2 - x - 2}{x+1}$ if x ≠ -1, and 1 if x = -1.

This function is discontinuous at x = -1 even though f(-1) is defined and the limit as x approaches -1 exists.

$$\lim_{x \to -1} f(x) = \frac{x^2 - x - 2}{x+1} = \lim_{x \to -1} \frac{(x-2)(x+1)}{x+1} = \lim_{x \to -1}(x-2) = -3$$

The function is discontinuous because $\lim_{x \to -1} f(x) \neq f(-1)$. The function does not satisfy all three necessary criteria.

Discontinuous functions have breaks in their graphs. In other words, we cannot sketch the graph of a discontinuous function without lifting the pen off the paper.

The second application of limits is determining asymptotes of a function's graph. If either $\lim_{x \to \infty} f(x) = L$ or $\lim_{x \to -\infty} f(x) = L$, then the line y = L is a horizontal asymptote of the graph of y = f(x). If either $\lim_{x \to a+} f(x) = \infty$ or $-\infty$ or $\lim_{x \to a-} f(x) = \infty$ or $-\infty$, then the line x = a is a vertical asymptote of the curve y = f(x). The graph of a function will approach, but never reach, an asymptote line.

The **difference quotient** is the average rate of change over an interval. For a function f, the **difference quotient** is represented by the formula:

$$\frac{f(x+h) - f(x)}{h}.$$

This formula computes the slope of the secant line through two points on the graph of f. These are the points with x-coordinates x and $x+h$.

Example: Find the difference quotient for the function
$f(x) = 2x^2 + 3x - 5$.

$$\frac{f(x+h) - f(x)}{h} = \frac{2(x+h)^2 + 3(x+h) - 5 - (2x^2 + 3x - 5)}{h}$$

$$= \frac{2(x^2 + 2hx + h^2) + 3x + 3h - 5 - 2x^2 - 3x + 5}{h}$$

$$= \frac{2x^2 + 4hx + 2h^2 + 3x + 3h - 5 - 2x^2 - 3x + 5}{h}$$

$$= \frac{4hx + 2h^2 + 3h}{h}$$

$$= 4x + 2h + 3$$

The **derivative** is the slope of a tangent line to a graph $f(x)$, and is usually denoted $f'(x)$. This is also referred to as the instantaneous rate of change.

The derivative of $f(x)$ at $x = a$ is given by taking the limit of the average rates of change (computed by the difference quotient) as h approaches 0.

$$f'(a) = \lim_{h \to 0} \frac{f(a+h) - f(a)}{h}$$

Example: Suppose a company's annual profit (in millions of dollars) is represented by the above function $f(x) = 2x^2 + 3x - 5$ and x represents the number of years in the interval. Compute the rate at which the annual profit was changing over a period of 2 years.

$$f'(a) = \lim_{h \to 0} \frac{f(a+h) - f(a)}{h}$$

$$= f'(2) = \lim_{h \to 0} \frac{f(2+h) - f(2)}{h}$$

Using the difference quotient we computed above, $4x + 2h + 3$, we get

$$f'(2) = \lim_{h \to 0}(4(2) + 2h + 3)$$
$$= 8 + 3$$
$$= 11.$$

We have, therefore, determined that the annual profit for the company has increased at the average rate of $11 million per year over the two-year period.

Exercise: Interpreting Slope as a Rate of Change

Connection: Social Sciences/Geography

Real-life Application: Slope is often used to describe a constant or average rate of change. These problems usually involve units of measure such as miles per hour or dollars per year.

Problem:

The town of Verdant Slopes has been experiencing a boom in population growth. By the year 2000, the population had grown to 45,000, and by 2005, the population had reached 60,000.

Communicating about Algebra:

a. Using the formula for slope as a model, find the average rate of change in population growth, expressing your answer in people per year.

Extension:

b. Using the average rate of change determined in a., predict the population of Verdant Slopes in the year 2010.

Solution:

a. *Let t represent the time and p represent population growth. The two observances are represented by (t_1, p_1) and (t_2, p_2).*

1st observance = (t_1, p_1) = (2000, 45000)
2nd observance = (t_2, p_2) = (2005, 60000)

Use the formula for slope to find the average rate of change.

$$\text{Rate of change} = \frac{p_2 - p_1}{t_2 - t_1}$$

Substitute values.

$$= \frac{60000 - 45000}{2005 - 2000}$$

Simplify.

$$= \frac{15000}{5} = 3000 \, people \, / \, year$$

The average rate of change in population growth for Verdant Slopes between the years 2000 and 2005 was 3000 people/year.

b.

$$3000 \, people \, / \, year \times 5 \, years = 15000 \, people$$
$$60000 \, people + 15000 \, people = 75000 \, people$$

At a continuing average rate of growth of 3000 people/year, the population of Verdant Slopes could be expected to reach 75,000 by the year 2010.

COMPETENCY 0018   **UNDERSTAND AND APPLY THE PRINCIPLES AND TECHNIQUES OF DIFFERENTIAL CALCULUS.**

The derivative of a function has two basic interpretations.

    I.   Instantaneous rate of change
    II.   Slope of a tangent line at a given point

If a question asks for the rate of change of a function, take the derivative to find the equation for the rate of change. Then plug in for the variable to find the instantaneous rate of change.

The following is a list summarizing some of the more common quantities referred to in rate of change problems.

area	height	profit
decay	population growth	sales
distance	position	temperature
frequency	pressure	volume

Pick a point, say $x = {}^-3$, on the graph of a function. Draw a tangent line at that point. Find the derivative of the function and plug in $x = {}^-3$. The result will be the slope of the tangent line.

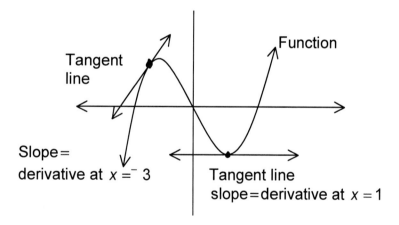

To find the **slope of a curve** at a point, there are two steps to follow.

    1.   Take the derivative of the function.
    2.   Plug in the value to find the slope.

If plugging into the derivative yields a value of zero, the tangent line is horizontal at that point.

If plugging into the derivative produces a fraction with zero in the denominator, the tangent line at this point has an undefined slope and is thus a vertical line.

Examples:

1. Find the slope of the tangent line for the given function at the given point.

   $$y = \frac{1}{x-2} \text{ at } (3,1)$$

   $$y = (x-2)^{-1}$$ Rewrite using negative exponents.

   $$\frac{dy}{dx} = {}^-1(x-2)^{-1-1}(1)$$ Use the Chain rule. The derivative of $(x-2)$ is 1.

   $$\frac{dy}{dx} = {}^-1(x-2)^{-2}$$

   $$\frac{dy}{dx}_{x=3} = {}^-1(3-2)^{-2}$$ Evaluate at $x=3$.

   $$\frac{dy}{dx}_{x=3} = {}^-1$$ The slope of the tangent line is $^-1$ at $x=3$.

2. Find the points where the tangent to the curve $f(x) = 2x^2 + 3x$ is parallel to the line $y = 11x - 5$.

   $$f'(x) = 2 \bullet 2x^{2-1} + 3$$ Take the derivative of $f(x)$ to get the slope of a tangent line.

   $$f'(x) = 4x + 3$$

   $$4x + 3 = 11$$ Set the slope expression $(4x+3)$ equal to the slope of $y = 11x - 5$.

   $$x = 2$$ Solve for the x value of the point.

   $$f(2) = 2(2)^2 + 3(2)$$ The y value is 14.

   $$f(2) = 14$$ So $(2,14)$ is the point on $f(x)$ where the tangent line is parallel to $y = 11x - 5$.

To **write an equation** of a tangent line at a point, two things are needed.

A point--the problem will usually provide a point, (x,y). If the problem only gives an x value, plug the value into the original function to get the y coordinate.

The slope--to find the slope, take the derivative of the original function. Then plug in the x value of the point to get the slope.

After obtaining a point and a slope, use the Point-Slope form for the equation of a line:

$$(y - y_1) = m(x - x_1)$$

where m is the slope and (x_1, y_1) is the point.

Example:

Find the equation of the tangent line to $f(x) = 2e^{x^2}$ at $x = {}^-1$.

$f({}^-1) = 2e^{({}^-1)^2}$	Plug in the x coordinate to obtain the y coordinate.
$= 2e^1$	The point is $({}^-1, 2e)$.
$f'(x) = 2e^{x^2} \bullet (2x)$	
$f'({}^-1) = 2e^{({}^-1)^2} \bullet (2 \bullet {}^-1)$	
$f'({}^-1) = 2e^1({}^-2)$	
$f'({}^-1) = {}^-4e$	The slope at $x = {}^-1$ is ${}^-4e$.
$(y - 2e) = {}^-4e(x - {}^-1)$	Plug in the point $({}^-1, 2e)$ and the slope $m = {}^-4e$. Use the point slope form of a line.
$y = {}^-4ex - 4e + 2e$	
$y = {}^-4ex - 2e$	Simplify to obtain the equation for the tangent line.

Algebraic Functions

A. Derivative of a constant--for any constant, the derivative is always zero.

B. Derivative of a variable--the derivative of a variable (i.e. x) is one.

C. Derivative of a variable raised to a power--for variable expressions with rational exponents (i.e. $3x^2$) multiply the coefficient (3) by the exponent (2) then subtract one (1) from the exponent to arrive at the derivative $\left(3x^2\right) = (6x)$

Example:

1. $y = 5x^4$ Take the derivative.

   $\dfrac{dy}{dx} = (5)(4)x^{4-1}$

   Multiply the coefficient by the exponent and subtract 1 from the exponent.

   $\dfrac{dy}{dx} = 20x^3$ Simplify.

2. $y = \dfrac{1}{4x^3}$ Rewrite using negative exponent.

   $y = \dfrac{1}{4}x^{-3}$ Take the derivative.

   $\dfrac{dy}{dx} = \left(\dfrac{1}{4} \bullet {}^{-}3\right)x^{-3-1}$

   $\dfrac{dy}{dx} = \dfrac{{}^{-}3}{4}x^{-4} = \dfrac{{}^{-}3}{4x^4}$ Simplify.

3. $y = 3\sqrt{x^5}$ Rewrite using $\sqrt[z]{x^n} = x^{n/z}$.

   $y = 3x^{5/2}$ Take the derivative.

   $\dfrac{dy}{dx} = (3)\left(\dfrac{5}{2}\right)x^{5/2-1}$

   $\dfrac{dy}{dx} = \left(\dfrac{15}{2}\right)x^{3/2}$ Simplify.

   $\dfrac{dy}{dx} = 7.5\sqrt{x^3} = 7.5x\sqrt{x}$

Trigonometric Functions

A. $\sin x$ --the derivative of the sine of x is simply the cosine of x.

B. $\cos x$ --the derivative of the cosine of x is negative one ($^{-}1$) times the sine of x.

C. $\tan x$ --the derivative of the tangent of x is the square of the secant of x.

If the object of the trig. function is an expression other than x, follow the above rules substituting the expression for x. The only additional step is to multiply the result by the derivative of the expression.

Examples:

1. $y = \pi \sin x$
    $$\frac{dy}{dx} = \pi \cos x$$

    Carry the coefficient (π) throughout the problem.

2. $y = \dfrac{2}{3}\cos x$

    $$\frac{dy}{dx} = \frac{^{-}2}{3}\sin x$$

    Do not forget to multiply the coefficient by negative one when taking the derivative of a cosine function.

3. $y = 4\tan\left(5x^3\right)$

    $$\frac{dy}{dx} = 4\sec^2\left(5x^3\right)\left(5 \bullet 3x^{3-1}\right)$$

    The derivative of $\tan x$ is $\sec^2 x$.

    $$\frac{dy}{dx} = 4\sec^2\left(5x^3\right)\left(15x^2\right)$$

    Carry the $\left(5x^3\right)$ term throughout the problem.

    $$\frac{dy}{dx} = 4 \bullet 15x^2 \sec^2\left(5x^3\right)$$

    Multiply $4\sec^2\left(5x^3\right)$ by the derivative of $5x^3$.

    $$\frac{dy}{dx} = 60x^2 \sec^2\left(5x^3\right)$$

    Rearrange the terms and simplify.

Exponential Functions

$f(x) = e^x$ is an exponential function. The derivative of e^x is exactly the same thing→ e^x. If instead of x, the exponent on e is an expression, the derivative is the same e raised to the algebraic exponent multiplied by the derivative of the algebraic expression.

If a base other than e is used, the derivative is the natural log of the base times the original exponential function times the derivative of the exponent.

Examples:

1. $y = e^x$

    $\dfrac{dy}{dx} = e^x$

2. $y = e^{3x}$

    $\dfrac{dy}{dx} = e^{3x} \bullet 3 = 3e^{3x}$ Multiply e^{3x} by the derivative of $3x$ which is 3.

    $\dfrac{dy}{dx} = 3e^{3x}$ Rearrange the terms.

3. $y = \dfrac{5}{e^{\sin x}}$

    $y = 5e^{-\sin x}$ Rewrite using negative exponents

    $\dfrac{dy}{dx} = 5e^{-\sin x} \bullet \left({}^{-}\cos x \right)$ Multiply $5e^{-\sin x}$ by the derivative of ${}^{-}\sin x$ which is ${}^{-}\cos x$.

    $\dfrac{dy}{dx} = \dfrac{{}^{-}5\cos x}{e^{\sin x}}$ Use the definition of negative exponents to simplify.

4. $y = {}^{-}2 \bullet \ln 3^{4x}$

    $\dfrac{dy}{dx} = {}^{-}2 \bullet (\ln 3)\left(3^{4x} \right)(4)$ The natural log of the base is ln3.

        The derivative of $4x$ is 4.

    $\dfrac{dy}{dx} = {}^{-}8 \bullet 3^{4x} \ln 3$ Rearrange terms to simplify.

Logarithmic Functions

The most common logarithmic function on the Exam is the natural logarithmic function ($\ln x$). The derivative of $\ln x$ is simply $1/x$. If x is replaced by an algebraic expression, the derivative is the fraction one divided by the expression multiplied by the derivative of the expression.

For all other logarithmic functions, the derivative is 1 over the argument of the logarithm multiplied by 1 over the natural logarithm (ln) of the base multiplied by the derivative of the argument.

Examples:

1. $y = \ln x$

$$\frac{dy}{dx} = \frac{1}{x}$$

2. $y = 3\ln\left(x^{-2}\right)$

$$\frac{dy}{dx} = 3 \bullet \frac{1}{x^{-2}} \bullet \left(^{-}2x^{-2-1}\right)$$

Multiply one over the argument (x^{-2}) by the derivative of x^{-2} which is $^{-}2x^{-2-1}$.

$$\frac{dy}{dx} = 3 \bullet x^2 \bullet \left(^{-}2x^{-3}\right)$$

$$\frac{dy}{dx} = \frac{^{-}6x^2}{x^3}$$

Simplify using the definition of negative exponents.

$$\frac{dy}{dx} = \frac{^{-}6}{x}$$

Cancel common factors to simplify.

3. $y = \log_5(\tan x)$

$$\frac{dy}{dx} = \frac{1}{\tan x} \bullet \frac{1}{\ln 5} \bullet (\sec^2 x)$$

The derivative of $\tan x$ is $\sec^2 x$.

$$\frac{dy}{dx} = \frac{\sec^2 x}{(\tan x)(\ln 5)}$$

Implicitly Defined Functions

Implicitly defined functions are ones where both variables (usually x and y) appear in the function. All of the rules for finding the derivative still apply to both variables. The only difference is that, while the derivative of x is one (1) and typically not even mentioned, the derivative of y must be written y' or dy/dx. Work these problems just like the other derivative problems, just remember to include the extra step of multiplying by dy/dx in the result.

If the question asks for the derivative of y, given an equation with x and y on both sides, find the derivative of each side. Then solve the new equation for dy/dx just as you would an algebra problem.

Examples:

1. $\dfrac{d}{dx}\left(y^3\right) = 3y^{3-1} \cdot \dfrac{dy}{dx}$

Recall the derivative of x^3 is $3x^{3-1}$. Follow the same rule, but also multiply by the derivative of y which

$\dfrac{d}{dx}(y^3) = 3y^2 \dfrac{dy}{dx}$ is dy/dx.

2. $\dfrac{d}{dx}(3\ln y) = 3 \cdot \dfrac{1}{y} \cdot \dfrac{dy}{dx}$

3. $\dfrac{d}{dx}(^{-}2\cos y) = ^{-}2(^{-}1\sin y)\dfrac{dy}{dx}$

Recall the derivative of $\cos x$ is $^{-}\sin x$.

$\dfrac{d}{dx}(^{-}2\cos y) = 2\sin y \dfrac{dy}{dx}$

4. $2y = e^{3x}$

Solve for after taking the derivative.

$2 \cdot \dfrac{dy}{dx} = e^{3x} \cdot 3$

The derivative of e^{3x} is $e^{3x} \cdot 3$.

$\dfrac{dy}{dx} = \dfrac{3}{2}e^{3x}$

Divide both sides by 2 to solve for the derivative dy/dx.

Sums, Products, and Quotients

A.   Derivative of a sum--find the derivative of each term separately and add the results.

B.   Derivative of a product--multiply the derivative of the first factor by the second factor and add to it the product of the first factor and the derivative of the second factor.

Remember the phrase "first times the derivative of the second plus the second times the derivative of the first."

C.   Derivative of a quotient--use the rule "bottom times the derivative of the top minus the top times the derivative of the bottom all divided by the bottom squared."

Examples:

1. $y = 3x^2 + 2\ln x + 5\sqrt{x}$ $\sqrt{x} = x^{1/2}$.

$$\frac{dy}{dx} = 6x^{2-1} + 2 \bullet \frac{1}{x} + 5 \bullet \frac{1}{2}x^{1/2-1}$$

$$\frac{dy}{dx} = 6x + \frac{2}{x} + \frac{5}{2} \bullet \frac{1}{\sqrt{x}} = \frac{12x^2 + 4 + 5\sqrt{x}}{2x}$$

$$= \frac{12x^2 + 5\sqrt{x} + 4}{2x}$$

$$x^{1/2-1} = x^{-1/2} = 1/\sqrt{x} .$$

2. $y = 4e^{x^2} \bullet \sin x$

$$\frac{dy}{dx} = 4(e^{x^2} \bullet \cos x + \sin x \bullet e^{x^2} \bullet 2x$$

The derivative of
e^{x^2} is $e^{x^2} \bullet 2$.

$$\frac{dy}{dx} = 4(e^{x^2} \cos x + 2xe^{x^2} \sin x)$$

$$\frac{dy}{dx} = 4e^{x^2} \cos x + 8xe^{x^2} \sin x$$

3. $y = \dfrac{\cos x}{x}$

$$\frac{dy}{dx} = \frac{x(^-\sin x) - \cos x \bullet 1}{x^2}$$

The derivative of x
is 1.
The derivative of
$\cos x$ is
$^-\sin x$.

$$\frac{dy}{dx} = \frac{^-x \sin x - \cos x}{x^2}$$

Composite Functions (Chain Rule)

A composite function is made up of two or more separate functions such as $\sin(\ln x)$ or $x^2 e^{3x}$. To find the derivatives of these composite functions requires two steps. First identify the predominant function in the problem. For example, in $\sin(\ln x)$) the major function is the sine function. In $x^2 e^{3x}$ the major function is a product of two expressions (x^2 and e^{3x}). Once the predominant function is identified, apply the appropriate differentiation rule. Be certain to include the step of taking the derivative of every part of the functions which comprise the composite function. Use parentheses as much as possible.

Examples:

1. $y = \sin(\ln x)$ The major function is a sine function.

$\dfrac{dy}{dx} = [\cos(\ln x)] \bullet \left[\dfrac{1}{x}\right]$ The derivative of $\sin x$ is $\cos x$.

The derivative of $\ln x$ is $1/x$.

2. $y = x^2 \bullet e^{3x}$ The major function is a product.

$\dfrac{dy}{dx} = x^2\left(e^{3x} \bullet 3\right) + e^{3x} \bullet 2x$

The derivative of a product is

"First $\dfrac{dy}{dx} = 3x^2 e^{3x} + 2xe^{3x}$

times the derivative of second plus the second times the derivative of the first."

3. $y = \tan^2\left(\dfrac{\ln x}{\cos x}\right)$

This function is made of several functions. The major function is a power function.

$\dfrac{dy}{dx} = \left[2\tan^{2-1}\left(\dfrac{\ln x}{\cos x}\right)\right]\left[\sec^2\left(\dfrac{\ln x}{\cos x}\right)\right]\left[\dfrac{d}{dx}\left(\dfrac{\ln x}{\cos x}\right)\right]$

The derivative of $\tan x$ is $\sec^2 x$. Hold off one more to take the derivative of $\ln x / \cos x$.

$\dfrac{dy}{dx} = \left[2\tan\left(\dfrac{\ln x}{\cos x}\right)\sec^2\left(\dfrac{\ln x}{\cos x}\right)\right]\left[\dfrac{(\cos x)(1/x) - \ln x(^-\sin x)}{\cos^2 x}\right]$

$\dfrac{dy}{dx} = \left[2\tan\left(\dfrac{\ln x}{\cos x}\right)\sec^2\left(\dfrac{\ln x}{\cos x}\right)\right]\left[\dfrac{(\cos x)(1/x) + \ln x(\sin x)}{\cos^2 x}\right]$

The derivative of a quotient is "Bottom times the derivative of the top minus the top times the derivative of the bottom all divided by the bottom squared."

Higher Order Derivatives

If a question simply asks for the derivative of a function, the question is asking for the first derivative. To find the second derivative of a function, take the derivative of the first derivative. To find the third derivative, take the derivative of the second derivative; and so on. All of the regular derivative rules still apply.

Examples:

1. Find the second derivative $\left(\dfrac{d^2y}{dx^2} \text{ or } y''\right)$ of the following function:

$$y = 5x^2$$

$\dfrac{dy}{dx} = 2 \bullet 5x^{2-1} = 10$ Take the first derivative.

$\dfrac{d^2y}{dx^2} = 10$ The derivative of $10x$ is 10.

2. Find the third derivative (y''') of $f(x) = 4x^{3/2}$:

$$y' = \left(4 \bullet \frac{3}{2}\right)x^{\left(\frac{3}{2}-1\right)} = 6x^{\frac{1}{2}}$$

$$y'' = \left(6 \bullet \frac{1}{2}\right)x^{\left(\frac{1}{2}-1\right)} = 3x^{-\frac{1}{2}}$$

$$y''' = \left[3 - \left(\frac{1}{2}\right)\right]x^{\left(-\frac{1}{2}-1\right)} = -\frac{3}{2}x^{-\frac{3}{2}}$$

Increasing or Decreasing

A function is said to be increasing if it is rising from left to right and decreasing if it is falling from left to right. Lines with positive slopes are increasing, and lines with negative slopes are decreasing. If the function in question is something other than a line, simply refer to the slopes of the tangent lines as the test for increasing or decreasing. Take the derivative of the function and plug in an x value to get the slope of the tangent line; a positive slope means the function is increasing and a negative slope means it is decreasing. If an interval for x values is given, just pick any point between the two values to substitute.

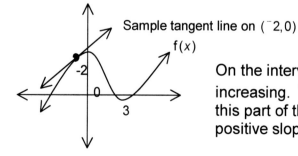

Sample tangent line on $(^-2,0)$

$f(x)$

On the interval $(^-2,0)$, $f(x)$ is increasing. The tangent lines on this part of the graph have positive slopes.

Example:

The growth of a certain bacteria is given by $f(x) = x + \dfrac{1}{x}$. Determine if the rate of growth is increasing or decreasing on the time interval $(^-1, 0)$.

$f'(x) = 1 + \dfrac{^-1}{x^2}$

To test for increasing or decreasing, find the slope of the tangent line by taking the derivative.

$f'\left(\dfrac{^-1}{2}\right) = 1 + \dfrac{^-1}{(^-1/2)^2}$

Pick any point on $(^-1, 0)$ and substitute into the derivative.

$f'\left(\dfrac{^-1}{2}\right) = 1 + \dfrac{^-1}{1/4}$

$= 1 - 4$

$= {}^-3$

The slope of the tangent line at $x = \dfrac{^-1}{2}$ is $^-3$. The exact value of the slope is not important. The important fact is that the slope is negative.

Maxima and Minima

Substituting an x value into a function produces a corresponding y value. The coordinates of the point (x, y), where y is the largest of all the y values, is said to be a **maximum point**. The coordinates of the point (x, y), where y is the smallest of all the y values, is said to be a **minimum point**. To find these points, only a few x values must be tested. First, find all of the x values that make the derivative either zero or undefined. Substitute these values into the original function to obtain the corresponding y values. Compare the y values. The largest y value is a maximum; the smallest y value is a minimum. If the question asks for the maxima or minima on an interval, be certain to also find the y values that correspond to the numbers at either end of the interval.

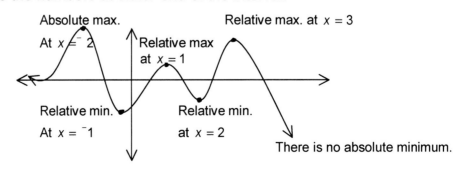

Absolute max.
At $x = {}^-2$

Relative max. at $x = 3$

Relative max at $x = 1$

Relative min.
At $x = {}^-1$

Relative min.
at $x = 2$

There is no absolute minimum.

Example:

Find the maxima and minima of $f(x) = 2x^4 - 4x^2$ at the interval $(^-2, 1)$.

$f'(x) = 8x^3 - 8x$	Take the derivative first. Find all the x values (critical values) that make the derivative zero or undefined. In this case, there are no x values that make the derivative undefined.
$8x^3 - 8x = 0$	
$8x(x^2 - 1) = 0$	
$8x(x - 1)(x + 1) = 0$	Substitute the critical values into the original function. Also, plug in the endpoint of the interval.
$x = 0,\ x = 1,\ \text{or } x =^- 1$	
$f(0) = 2(0)^4 - 4(0)^2 = 0$	
$f(1) = 2(1)^4 - 4(1)^2 =^- 2$	Note that 1 is a critical point
$f(^-1) = 2(^-1)^4 - 4(^-1)^2 =^- 2$	and an endpoint.
$f(^-2) = 2(^-2)^4 - 4(^-2)^2 = 16$	

The maximum is at (–2, 16) and there are minima at (1, –2) and (–1, –2). (0,0) is neither the maximum or minimum on (–2, 1) but it is still considered a relative extra point.

Concavity

The first derivative reveals whether a curve is rising or falling **(increasing or decreasing)** from the left to the right. In much the same way, the second derivative relates whether the curve is concave up or concave down. Curves which are concave up are said to "collect water;" curves which are concave down are said to "dump water." To find the intervals where a curve is concave up or concave down, follow the following steps.

1. Take the second derivative (i.e. the derivative of the first derivative).
2. Find the critical x values.
    -Set the second derivative equal to zero and solve for critical x values.
    -Find the x values that make the second derivative undefined (i.e. make the denominator of the second derivative equal to zero).
    Such values may not always exist.
3. Pick sample values which are both less than and greater than each of the critical values.

4. Substitute each of these sample values into the second derivative and determine whether the result is positive or negative.

-If the sample value yields a positive number for the second derivative, the curve is concave up on the interval where the sample value originated.

-If the sample value yields a negative number for the second derivative, the curve is concave down on the interval where the sample value originated.

Example:

Find the intervals where the curve is concave up and concave down for $f(x) = x^4 - 4x^3 + 16x - 16$.

$f'(x) = 4x^3 - 12x^2 + 16$ Take the second derivative.

$f''(x) = 12x^2 - 24x$

Find the critical values by setting the second derivative equal to zero.

$12x^2 - 24x = 0$

$12x(x - 2) = 0$

$x = 0$ or $x = 2$

There are no values that make the second derivative undefined. Set up a number line with the critical values.

Sample values: ⁻1, 1, 3

$f''(^-1) = 12(^-1)^2 - 24(^-1) = 36$

$f''(1) = 12(1)^2 - 24(1) = ^-12$

$f''(3) = 12(3)^2 - 24(3) = 36$

Pick sample values in each of the 3 intervals. If the sample value produces a negative number, the function is concave down.

If the value produces a positive number, the curve is concave up. If the value produces a zero, the function is linear.

Therefore when $x < 0$ the function is concave up,
when $0 < x < 2$ the function is concave down,
when $x > 2$ the function is concave up.

Points of Inflection

A point of inflection is a point where a curve changes from being concave up to concave down or vice versa. To find these points, follow the steps for finding the intervals where a curve is concave up or concave down. A critical value is part of an inflection point if the curve is concave up on one side of the value and concave down on the other. The critical value is the x coordinate of the inflection point. To get the y coordinate, plug the critical value into the original function.

Example: Find the inflection points of $f(x) = 2x - \tan x$ where $\dfrac{^-\pi}{2} < x < \dfrac{\pi}{2}$.

$(x) = 2x - \tan x \qquad \dfrac{^-\pi}{2} < x < \dfrac{\pi}{2}$
Note the restriction on x.

$f'(x) = 2 - \sec^2 x$
Take the second derivative.

$f''(x) = 0 - 2 \bullet \sec x \bullet (\sec x \tan x)$
Use the Power rule.

$= {}^-2 \bullet \dfrac{1}{\cos x} \bullet \dfrac{1}{\cos x} \bullet \dfrac{\sin x}{\cos x}$
The derivative of $\sec x$ is $(\sec x \tan x)$.

$f''(x) = \dfrac{^-2\sin x}{\cos^3 x}$
Find critical values by solving for the second derivative equal to zero.

$0 = \dfrac{^-2\sin x}{\cos^3 x}$
No x values on $\left(\dfrac{^-\pi}{2}, \dfrac{\pi}{2}\right)$ make the denominator zero.

${}^-2\sin x = 0$
$\sin x = 0$
$x = 0$
Pick sample values on each side of the critical value $x = 0$.

Sample values: $x = \dfrac{^-\pi}{4}$ and $x = \dfrac{\pi}{4}$

$$f'' \left(\dfrac{^-\pi}{4} \right) = \dfrac{^-2\sin(^-\pi/4)}{\cos^3(\pi/4)} = \dfrac{^-2(^-\sqrt{2}/2)}{(\sqrt{2}/2)^3} = \dfrac{\sqrt{2}}{(\sqrt{8}/8)} = \dfrac{8\sqrt{2}}{\sqrt{8}} = \dfrac{8\sqrt{2}}{\sqrt{8}} \cdot \dfrac{\sqrt{8}}{\sqrt{8}}$$

$$= \dfrac{8\sqrt{16}}{8} = 4$$

$$f'' \left(\dfrac{\pi}{4} \right) = \dfrac{^-2\sin(\pi/4)}{\cos^3(\pi/4)} = \dfrac{^-2(\sqrt{2}/2)}{(\sqrt{2}/2)^3} = \dfrac{^-\sqrt{2}}{(\sqrt{8}/8)} = \dfrac{^-8\sqrt{2}}{\sqrt{8}} = -4$$

The second derivative is positive on ($0, \infty$) and negative on ($^-\infty, 0$). So the curve changes concavity at $x = 0$. Use the original equation to find the y value that inflection occurs at.

$$f(0) = 2(0) - \tan 0 = 0 - 0 = 0 \qquad \text{The inflection point is (0,0).}$$

Extreme Value Problems

Extreme value problems are also known as max-min problems. Extreme value problems require using the first derivative to find values which either maximize or minimize some quantity such as area, profit, or volume. Follow these steps to solve an extreme value problem.

1. Write an equation for the quantity to be maximized or minimized.
2. Use the other information in the problem to write secondary equations.
3. Use the secondary equations for substitutions, and rewrite the original equation in terms of only one variable.
4. Find the derivative of the primary equation (step 1) and the critical values of this derivative.
5. Substitute these critical values into the primary equation.

The value which produces either the largest or smallest value is used to find the solution.

Example:

A manufacturer wishes to construct an open box from the piece of metal shown below by cutting squares from each corner and folding up the sides. The square piece of metal is 12 feet on a side. What are the dimensions of the squares to be cut out which will maximize the volume?

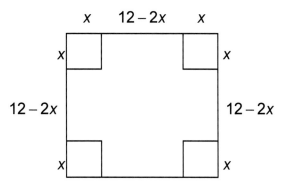

$Volume = lwh$	Primary equation.
$l = 12 - 2x$	
$w = 12 - 2x$	Secondary equations.
$h = x$	
$V = (12 - 2x)(12 - 2x)(x)$	Make substitutions.
$V = (144x - 48x^2 + 4x^3)$	Take the derivative.
$\dfrac{dV}{dx} = 144 - 96x + 12x^2$	
	Find critical values by setting the
$0 = 12(x^2 - 8x + 12)$	derivative equal to zero.
$0 = 12(x - 6)(x - 2)$	
$x = 6$ and $x = 2$	Substitute critical values into volume equation.

$$V = 144(6) - 48(6)^2 + 4(6)^3 \qquad V = 144(2) - 48(2)^2 + 4(2)^3$$

$$V = 0 \text{ ft}^3 \text{ when } x = 6 \qquad\qquad V = 128 \text{ ft}^3 \text{ when } x = 2$$

Therefore, the manufacturer can maximize the volume if the squares to be cut out are 2 feet by 2 feet ($x = 2$).

Velocity and Acceleration Problems

If a **particle** (or a car, a bullet, etc.) is moving along a line, then the distance that the particle travels can be expressed by a function in terms of time.

1. The first derivative of the distance function will provide the velocity function for the particle. Substituting a value for time into this expression will provide the instantaneous velocity of the particle at the time. Velocity is the rate of change of the distance traveled by the particle. Taking the absolute value of the derivative provides the speed of the particle. A positive value for the velocity indicates that the particle is moving forward, and a negative value indicates the particle is moving backwards.

2. The second derivative of the distance function (which would also be the first derivative of the velocity function) provides the acceleration function. The acceleration of the particle is the rate of change of the velocity. If a value for time produces a positive acceleration, the particle is speeding up; if it produces a negative value, the particle is slowing down. If the acceleration is zero, the particle is moving at a constant speed.

To find the time when a particle stops, set the first derivative (i.e. the velocity function) equal to zero and solve for time. This time value is also the instant when the particle changes direction.

Example:

The motion of a particle moving along a line is according to the equation:

$s(t) = 20 + 3t - 5t^2$ where s is in meters and t is in seconds. Find the position, velocity, and acceleration of a particle at $t = 2$ seconds.

$s(2) = 20 + 3(2) - 5(2)^2$ $\quad = 6$ meters	Plug $t = 2$ into the original equation to find the position.
$s\,'(t) = v(t) = 3 - 10t$	The derivative of the first function gives the velocity.
$v(2) = 3 - 10(2) = {}^-17\,\text{m/s}$	Plug $t = 2$ into the velocity function to find the velocity. ${}^-17\,\text{m/s}$ indicates the particle is moving backwards.
$s\,''(t) = a(t) = {}^-10$	The second derivation of position gives the acceleration. Substitute
$a(2) = {}^-10\,\text{m/s}^2$	$t = 2$, yields an acceleration of ${}^-10\,\text{m/s}^2$, which indicates the particle is slowing down.

COMPETENCY 0019   **UNDERSTAND AND APPLY THE PRINCIPLES AND TECHNIQUES OF INTEGRAL CALCULUS.**

Taking the integral of a function and evaluating it from one x value to another provides the **total area under the curve** (i.e. between the curve and the x axis). Remember, though, that regions above the x axis have "positive" area and regions below the x axis have "negative" area. You must account for these positive and negative values when finding the area under curves. Follow these steps.

1. Determine the x values that will serve as the left and right boundaries of the region.
2. Find all x values between the boundaries that are either solutions to the function or are values which are not in the domain of the function. These numbers are the interval numbers.
3. Integrate the function.
4. Evaluate the integral once for each of the intervals using the boundary numbers.
5. If any of the intervals evaluates to a negative number, make it positive (the negative simply tells you that the region is below the x axis).
6. Add the value of each integral to arrive at the area under the curve.

Example:

Find the area under the following function on the given intervals.
$f(x) = \sin x$; $(0, 2\pi)$

$\sin x = 0$ Find any roots to f(x) on $(0, 2\pi)$.
$x = \pi$

$(0, \pi)$ $(\pi, 2\pi)$ Determine the intervals using the boundary numbers and the roots.

$\int \sin x\, dx = {}^{-}\cos x$ Integrate f(x). We can ignore the constant c because we have numbers to use to evaluate the integral.

$$ {}^{-}\cos x \Big]_{x=0}^{x=\pi} = {}^{-}\cos \pi - ({}^{-}\cos 0) $$

$$ {}^{-}\cos x \Big]_{x=0}^{x=\pi} = {}^{-}(-1) + (1) = 2 $$

$$ {}^{-}\cos x \Big]_{x=\pi}^{x=2\pi} = {}^{-}\cos 2\pi - ({}^{-}\cos \pi) $$

$$-\cos x\Big]_{x=\pi}^{x=2\pi} = {}^-1 + ({}^-1) = {}^-2$$

The $^-2$ means that for $(\pi, 2\pi)$, the region is below the x axis, but the area is still 2.

$$\text{Area} = 2 + 2 = 4$$

Add the 2 integrals together to get the area.

Taking the **antiderivative of a function** is the opposite of taking the derivative of the function--much in the same way that squaring an expression is the opposite of taking the square root of the expression. For example, since the derivative of x^2 is $2x$ then the antiderivative of 2x is x^2. The key to being able to take antiderivatives is being as familiar as possible with the derivative rules.

To take the antiderivative of an algebraic function (the sum of products of coefficients and variables raised to powers other than negative one), take the antiderivative of each term in the function by following these steps.

1. Take the antiderivative of each term separately.
2. The coefficient of the variable should be equal to one plus the exponent.
3. If the coefficient is not one more than the exponent, put the correct coefficient on the variable and also multiply by the reciprocal of the number put in.

    Ex. For $4x^5$, the coefficient should be 6 not 4. So put in 6 and the reciprocal 1/6 to achieve $(4/6)6x^5$.

4. Finally take the antiderivative by replacing the coefficient and variable with just the variable raised to one plus the original exponent.

    Ex. For $(4/6)6x^5$, the antiderivative is $(4/6)x^6$.

    You have to add in constant c because there is no way to know if a constant was originally present since the derivative of a constant is zero.

5. Check your work by taking the first derivative of your answer. You should get the original algebraic function.

Examples: Take the antiderivative of each function.

1. $f(x) = 5x^4 + 2x$ The coefficient of each term is already 1 more than the exponent.

$F(x) = x^5 + x^2 + c$ $F(x)$ is the antiderivative of $f(x)$

$F'(x) = 5x^4 + 2x$ Check by taking the derivative of $F(x)$.

2. $f(x) = {}^-2x^{\,-3}$

$F(x) = x^{\,-2} + c = \dfrac{1}{x^2} + c$ $F(x)$ is the antiderivative of $f(x)$.

$F'(x) = {}^-2x^{\,-3}$ Check.

3. $f(x) = {}^-4x^2 + 2x^7$ Neither coefficient is correct.

$f(x) = {}^-4 \bullet \dfrac{1}{3} \bullet 3x^2 + 2 \bullet \dfrac{1}{8} \bullet 8x^7$

Put in the correct coefficient along with its reciprocal.

$F(x) = {}^-4 \bullet \dfrac{1}{3}x^3 + 2 \bullet \dfrac{1}{8}x^8 + c$ $F(x)$ is the antiderivative of $f(x)$.

$F(x) = \dfrac{{}^-4}{3}x^3 + \dfrac{1}{4}x^8 + c$

$F'(x) = {}^-4x^2 + 2x^7$ Check.

The rules for taking antiderivatives of trigonometric functions follow, but be aware that these can get very confusing to memorize because they are very similar to the derivative rules. Check the antiderivative you get by taking the derivative and comparing it to the original function.

1. $\sin x$ the antiderivative for $\sin x$ is ${}^-\cos x + c$.
2. $\cos x$ the antiderivative for $\cos x$ is $\sin x + c$.
3. $\tan x$ the antiderivative for $\tan x$ is $-\ln|\cos x| + c$.
4. $\sec^2 x$ the antiderivative for $\sec^2 x$ is $\tan x + c$.
5. $\sec x \tan x$ the antiderivative for $\sec x \tan x$ is $\sec x + c$.

If the trigonometric function has a coefficient, simply keep the coefficient and multiply the antiderivative by it.

Examples: Find the antiderivatives for the following functions.

1. $f(x) = 2\sin x$ — Carry the 2 throughout the problem.

$F(x) = 2(^-\cos x) = {}^-2\cos x + c$ — $F(x)$ is the antiderivative.

$F'(x) = {}^-2(^-\sin x) = 2\sin x$ — Check by taking the derivative of $F(x)$.

2. $f(x) = \dfrac{\tan x}{5}$

$F(x) = \dfrac{-\ln|\cos x|}{5} + c$ — $F(x)$ is the antiderivative of $f(x)$.

$F'(x) = \dfrac{1}{5}\left(-\dfrac{1}{|\cos x|}\right)(-\sin x) = \dfrac{1}{5}\tan x$ — Check by taking the derivative of $F(x)$.

Practice problems: Find the antiderivative of each function.

1. $f(x) = {}^-20\cos x$
2. $f(x) = \pi \sec x \tan x$

Use the following rules when finding the antiderivative of an exponential function.

1. e^x — The antiderivative of e^x is the same $e^x + c$.
2. a^x — The antiderivative of a^x, where a is any number, is $a^x/\ln a + c$.

Examples: Find the antiderivatives of the following functions:

1. $f(x) = 10e^x$

$F(x) = 10e^x + c$ — $F(x)$ is the antiderivative.

$F'(x) = 10e^x$ — Check by taking the derivative of $F(x)$.

2. $F(x) = \dfrac{2^x}{3}$

$F(x) = \dfrac{1}{3} \cdot \dfrac{2^x}{\ln 2} + c$ — $F(x)$ is the antiderivative.

$F'(x) = \dfrac{1}{3\ln 2}\ln 2 (2^x)$ — Check by taking the derivative of $F(x)$.

$F'(x) = \dfrac{2^x}{3}$

An integral is almost the same thing as an antiderivative, the only difference is the notation.

$\int_{-2}^{1} 2x \, dx$ is the integral form of the antiderivative of $2x$. The numbers at the top and bottom of the integral sign (1 and $^-2$) are the numbers used to find the exact value of this integral. If these numbers are used the integral is said to be *definite* and does not have an unknown constant c in the answer.

The fundamental theorem of calculus states that an integral such as the one above is equal to the antiderivative of the function inside (here $2x$) evaluated from $x = {}^-2$ to $x = 1$. To do this, follow these steps.

1. Take the antiderivative of the function inside the integral.

2. Plug in the upper number (here $x = 1$) and plug in the lower number (here $x = {}^- 2$), giving two expressions.

3. Subtract the second expression from the first to achieve the integral value.

Examples:

1. $\int_{-2}^{1} 2x \, dx = x^2 \Big]_{-2}^{1}$ Take the antiderivative.

    $\int_{-2}^{1} 2x \, dx = 1^2 - (^-2)^2$ Substitute in $x = 1$ and $x = {}^-2$ and subtract the results.

    $\int_{-2}^{1} 2x \, dx = 1 - 4 = {}^- 3$ The integral has the value $^-3$.

2. $\int_{0}^{\pi/2} \cos x \, dx = \sin x \Big]_{0}^{\pi/2}$

                                  The antiderivative of $\cos x$ is $\sin x$.

    $\int_{0}^{\frac{\pi}{2}} \cos x \, dx = \sin \frac{\pi}{2} - \sin 0$ Substitute in $x = \frac{\pi}{2}$ and $x = 0$.

                                    Subtract the results.

    $\int_{0}^{\frac{\pi}{2}} \cos x \, dx = 1 - 0 = 1$ The integral has the value 1.

A list of integration formulas follows. In each case the letter u is used to represent either a single variable or an expression. Note that also in each case du is required. du is the derivative of whatever u stands for. If u is sin x then du is cosx, which is the derivative of sinx. If the derivative of u is not entirely present, remember you can put in constants as long as you also insert the reciprocal of any such constants. n is a natural number.

$$\int u^n du = \frac{1}{n+1} u^{n+1} + c \quad \text{if} \ n \neq {}^-1$$

$$\int \frac{1}{u} du = \ln|u| + c$$

$$\int e^u du = e^u + c$$

$$\int a^u du = \frac{1}{\ln a} a^u + c$$

$$\int \sin u\, du = {}^-\cos u + c$$

$$\int \cos u\, du = \sin u + c$$

$$\int \sec^2 u\, du = \tan u + c$$

$$\int \csc^2 u\, du = {}^-\cot u + c$$

Example:

1. $\displaystyle\int \frac{6}{x} dx = 6 \int \frac{1}{x} dx$

You can pull any constants outside the integral.

$$\int \frac{6}{x} dx = 6\ln|x| + c$$

Sometimes an integral does not always look exactly like the forms from above. But with a simple substitution (sometimes called a u substitution), the integral can be made to look like one of the general forms.

You might need to experiment with different *u* substitutions before you find the one that works. Follow these steps.

1. If the object of the integral is a sum or difference, first split the integral up.
2. For each integral, see if it fits one of the general forms from the previous pg.
3. If the integral does not fit one of the forms, substitute the letter *u* in place of one of the expressions in the integral.
4. Off to the side, take the derivative of *u*, and see if that derivative exists inside the original integral. If it does, replace that derivative expression with *du*. If it does not, try another *u* substitution.
5. Now the integral should match one of the general forms, including both the *u* and the *du*.
6. Take the integral using the general forms, and substitute for the value of *u*.

Examples:

1. $\int \left(\sin x^2 \bullet 2x + \cos x^2 \bullet 2x \right) dx$ Split the integral up.

$\int \sin(x^2) \bullet 2x\,dx + \int \cos(x^2) 2x\,dx$

$u = x^2, \quad du = 2x\,dx$ If you let $u = x^2$, the derivative of *u*, *du*, is $2x\,dx$.

$\int \sin u\,du + \int \cos u\,du$ Make the *u* and *du* substitutions.

$^{-}\cos u + \sin u + c$ Use the formula for integrating $\sin u$ and $\cos u$.

$^{-}\cos(x^2) + \sin(x^2) + c$ Substitute back in for *u*.

2. $\int e^{\sin x} \cos x\,dx$ Try letting $u = \cos x$. The derivative of *u* would be

$u = \cos x, \quad du = ^{-}\sin x\,dx$ $^{-}\sin x\,dx$, which is not present.

$u = \sin x, \quad du = \cos x\,dx$ Try another substitution: $u = \sin x, du = \cos x\,dx$. $du = \cos x\,dx$ is present.

$\int e^u\,du$ e^u is one of the general forms.

e^u The integral of e^u is e^u.

$e^{\sin x}$ Substitute back in for *u*.

Integration by parts should only be used if none of the other integration methods works. Integration by parts requires two substitutions (both *u* and *dv*).

1. Let *dv* be the part of the integral that you think can be integrated by itself.
2. Let *u* be the part of the integral which is left after the *dv* substitution is made.
3. Integrate the *dv* expression to arrive at just simply *v*.
4. Differentiate the *u* expression to arrive at *du*. If *u* is just *x*, then *du* is dx.
5. Rewrite the integral using $\int u\,dv = uv - \int v\,du$.
6. All that is left is to integrate $\int v\,du$.
7. If you cannot integrate *v du,* try a different set of substitutions and start the process over.

Examples:

1. $\int xe^{3x}\,dx$

Make *dv* and *u* substitutions.

$dv = e^{3x}\,dx \qquad u = x$

Integrate the *dv* term to arrive at *v.*

$v = \frac{1}{3}e^{3x} \qquad du = dx$

Differentiate the *u* term to arrive at *du.*

$\int xe^{3x}\,dx = x\left(\frac{1}{3}e^{3x}\right) - \int \frac{1}{3}e^{3x}\,dx$

Rewrite the integral using the above formula. Before taking the integral

$\int xe^{3x}\,dx = \frac{1}{3}xe^{3x} - \frac{1}{3} \cdot \frac{1}{3}\int e^{3x}3\,dx$

of $\frac{1}{3}e^{3x}\,dx$, you must put in a 3 and another 1/3.

$\int xe^{3x}\,dx = \frac{1}{3}xe^{3x} - \frac{1}{9}e^{3x} + c$

Integrate to arrive at the solution.

2. $\int \ln 4x \, dx$

Note that no other integration method will work.

$dv = dx \qquad u = \ln 4x$

Make the dv and u substitutions.

$v = x \qquad du = \dfrac{1}{4x} \cdot 4 = \dfrac{1}{x} dx$

Integrate dx to get x.

Differentiate ln4x to get $(1/x)\,dx$.

$\int \ln 4x \, dx = \ln 4x \cdot x - \int x \cdot \dfrac{1}{x} dx$

Rewrite the formula above.

$\int \ln 4x \, dx = \ln 4x \cdot x - \int dx$

Simplify the integral.

$\int \ln 4 \, dx = \ln 4x \cdot x - x + c$

Integrate dx to get the value $x + c$

Velocity Problems

The derivative of a distance function provides a velocity function, and the derivative of a velocity function provides an acceleration function. Therefore taking the antiderivative of an acceleration function yields the velocity function, and the antiderivative of the velocity function yields the distance function.

Example:

A particle moves along the x axis with acceleration $a(t) = 3t - 1$ cm/sec/sec. At time $t = 4$, the particle is moving to the left at 3 cm per second. Find the velocity of the particle at time $t = 2$ seconds.

$a(t) = 3t - 1$

Before taking the antiderivative, make sure the correct coefficients are present.

$a(t) = 3 \cdot \dfrac{1}{2} \cdot 2t - 1$

$v(t) = \dfrac{3}{2}t^2 - 1 \cdot t + c$

$v(t)$ is the antiderivative of $a(t)$.

$v(4) = \dfrac{3}{2}(4)^2 - 1(4) + c = {}^{-}3$

Use the given information that $v(4) = {}^{-}3$ to find c.

$$24 - 4 + c = {}^- 3$$
$$20 + c = {}^- 3$$
$$c = {}^- 23$$ The constant is $^-23$.

$$v(t) = \frac{3}{2} t^2 - 1t + {}^- 23$$ Rewrite $v(t)$ using $c = {}^- 23$.

$$v(2) = \frac{3}{2} 2^2 - 1(2) + {}^- 23$$ Solve $v(t)$ at t=2.

$$v(2) = 6 + 2 + {}^- 23 = {}^- 15$$ the velocity at $t = 2$ is -15 cm/sec.

Practice problem:

A particle moves along a line with acceleration $a(t) = 5t + 2$. The velocity after 2 seconds is $^-10$ m/sec.

1. Find the initial velocity.
2. Find the velocity at $t = 4$.

Distance Problems

To find the distance function, take the antiderivative of the velocity function. And to find the velocity function, find the antiderivative of the acceleration function. Use the information in the problem to solve for the constants that result from taking the antiderivatives.

Example:

A particle moves along the x axis with acceleration $a(t) = 6t - 6$. The initial velocity is 0 m/sec and the initial position is 8 cm to the right of the origin. Find the velocity and position functions.

$$v(0) = 0$$ Interpret the given information.
$$s(0) = 8$$
$$a(t) = 6t - 6$$ Put in the coefficients needed to take the antiderivative.

$$a(t) = 6 \bullet \frac{1}{2} \bullet 2t - 6$$

$$v(t) = \frac{6}{2} t^2 - 6t + c$$

Take the antiderivative of $a^{(t)}$ to get $v(t)$.

$$v(0) = 3(0)^2 - 6(0) + c = 0$$ Use $v(0) = 0$ to solve for c.
$$0 - 0 + c = 0$$
$$c = 0$$ $$c = 0$$

$v(t) = 3t^2 - 6t + 0$ Rewrite $v(t)$ using $c = 0$.

$v(t) = 3t^2 - 6\dfrac{1}{2} \bullet 2t$

Put in the coefficients needed to take the antiderivative.

$s(t) = t^3 - \dfrac{6}{2}t^2 + c$

Take the antiderivative of $v^{(t)}$ to get $s(t) \rightarrow$ the distance function.

$s(0) = 0^3 - 3(0)^2 + c = 8$ Use $s(0) = 8$ to solve for c.

$c = 8$

$s(t) = t^3 - 3t^2 + 8$

Area Between Two Curves

Finding the area between two curves is much the same as finding the area under one curve. But instead of finding the roots of the functions, you need to find the x values which produce the same number from both functions (set the functions equal and solve). Use these numbers and the given boundaries to write the intervals.
On each interval you must pick sample values to determine which function is "on top" of the other. Find the integral of each function. For each interval, subtract the "bottom" integral from the "top" integral. Use the interval numbers to evaluate each of these differences. Add the evaluated integrals to get the total area between the curves.

Example:

Find the area of the regions bounded by the two functions on the indicated intervals.

$$ff(x) = x + 2 \quad \text{and} \quad g(x) = x^2 \quad [-2,3]$$

$x + 2 = x^2$

Set the functions equal and solve.

$0 = (x - 2)(x + 1)$

$x = 2$ or $x = {}^{-}1$

$({}^{-}2, {}^{-}1) \; ({}^{-}1, 2) \; (2, 3)$

Use the solutions and the boundary numbers to write the intervals.

$$f(^-3/2) = \left(\frac{^-3}{2}\right) + 2 = \frac{1}{2}$$

Pick sample values on the integral and evaluate each function as that number.

$$g(^-3/2) = \left(\frac{^-3}{2}\right)^2 = \frac{9}{4}$$

$g(x)$ is "on top" on $\left[^-2, ^-1\right]$.

$$f(0) = 2$$
$$g(0) = 0$$

$f(x)$ is "on top" on $\left[^-1, 2\right]$.

$$f(5/2) = \frac{5}{2} + 2 = \frac{9}{2}$$

$g(x)$ is "on top" on $[2,3]$.

$$g(5/2) = \left(\frac{5}{2}\right)^2 = \frac{25}{4}$$

$$\int f(x)dx = \int (x + 2)dx$$

$$\int f(x)dx = \int xdx + 2\int dx$$

$$\int f(x)dx = \frac{1}{1+1}x^{1+1} + 2x$$

$$\int f(x)dx = \frac{1}{2}x^2 + 2x$$

$$\int g(x)dx = \int x^2 dx$$

$$\int g(x)dx = \frac{1}{2+1}x^{2+1} = \frac{1}{3}x^3$$

$$\text{Area } 1 = \int g(x)dx - \int f(x)dx \qquad g(x) \text{ is "on top" on } \left[^-2, ^-1\right].$$

$$\text{Area } 1 = \frac{1}{3}x^3 - \left(\frac{1}{2}x^2 + 2x\right)\Big]_{^-2}^{^-1}$$

$$\text{Area } 1 = \left[\frac{1}{3}(^-1)^3 - \left(\frac{1}{2}(^-1)^2 + 2(^-1)\right)\right] - \left[\frac{1}{3}(^-2)^3 - \left(\frac{1}{2}(^-2)^2 + 2(^-2)\right)\right]$$

$$\text{Area } 1 = \left[\frac{^-1}{3} - \left(\frac{^-3}{2}\right)\right] - \left[\frac{^-8}{3} - (^-2)\right]$$

$$\text{Area } 1 = \left(\frac{7}{6}\right) - \left(\frac{^-2}{3}\right) = \frac{11}{6}$$

$$\text{Area } 2 = \int f(x)dx - \int g(x)dx$$

f(x) is "on top" on $\left[^-1,2\right]$.

$$\text{Area } 2 = \frac{1}{2}x^2 + 2x - \frac{1}{3}x^3 \Big]_{^-1}^{2}$$

$$\text{Area } 2 = \left(\frac{1}{2}(2)^2 + 2(2) - \frac{1}{3}(2)^3\right) - \left(\frac{1}{2}(^-1)^2 + 2(^-1) - \frac{1}{3}(^-1)^3\right)$$

$$\text{Area } 2 = \left(\frac{10}{3}\right) - \left(\frac{1}{2} - 2 + \frac{1}{3}\right)$$

$$\text{Area } 2 = \frac{27}{6}$$

$$\text{Area } 3 = \int g(x)dx - \int f(x)dx$$

g(x) is "on top" on [2,3].

$$\text{Area } 3 = \frac{1}{3}x^3 - \left(\frac{1}{2}x^2 + 2x\right)\Big]_{2}^{3}$$

$$\text{Area } 3 = \left[\frac{1}{3}(3)^3 - \left(\frac{1}{2}(3^2) + 2(3)\right)\right] - \left[\frac{1}{3}(2)^3 - \left(\frac{1}{2}(2)^2 + 2(2)\right)\right]$$

$$\text{Area } 3 = \left(\frac{^-3}{2}\right) - \left(\frac{^-10}{3}\right) = \frac{11}{6}$$

$$\text{Total area} = \frac{11}{6} + \frac{27}{6} + \frac{11}{6} = \frac{49}{6} = 8\frac{1}{6}$$

Volume Problems

If you take the area bounded by a curve or curves and revolve it about a line, the result is a solid of revolution. To find the volume of such a solid, the Washer Method works in most instances. Imagine slicing through the solid perpendicular to the line of revolution. The "slice" should resemble a washer. Use an integral and the formula for the volume of disk.

$$Volume_{disk} = \pi \bullet radius^2 \bullet thickness$$

Depending on the situation, the radius is the distance from the line of revolution to the curve; or if there are two curves involved, the radius is the difference between the two functions. The thickness is *dx* if the line of revolution is parallel to the *x* axis and *dy* if the line of revolution is parallel to the *y* axis. Finally, integrate the volume expression using the boundary numbers from the interval.

Example:

Find the value of the solid of revolution found by revolving $f(x) = 9 - x^2$ about the x axis on the interval $[0,4]$.

radius $= 9 - x^2$
thickness $= dx$

Volume $= \int_0^4 \pi(9 - x^2)^2 \, dx$

Use the formula for volume of a disk.

Volume $= \pi \int_0^4 \left(81 - 18x^2 + x^4\right) dx$

Volume $= \pi \left(81x - \dfrac{18}{2+1}x^3 + \dfrac{1}{4+1}x^5 \right) \Big]_0^4$

Take the integral.

Volume $= \pi \left(81x - 6x^3 + \dfrac{1}{5}x^5 \right) \Big]_0^4$

Evaluate the integral first and subtract. $x = 4$ then at $x = 0$

Volume $= \pi \left[\left(324 - 384 + \dfrac{1024}{5} \right) - (0 - 0 + 0) \right]$

Volume $= \pi \left(144\dfrac{4}{5} \right) = 144\dfrac{4}{5}\pi = 454.9$

In a **differential equation**, the variables are derivatives of functions. First-order differential equations contain only first derivatives. Differential equations arise in the fundamental laws of economics, physics, chemistry, and biology. The general form of a first-order differential equation is

$$\frac{dy}{dx} = F(x, y)$$

where F is a function of two variables x and y. An important method of solving differential equations is separation of variables. When we can factor the function F into two functions, $g(x)$ and $f(y)$, the equation is separable. We solve such equations as follows.

$$\frac{dy}{dx} = g(x)f(y)$$

$$\frac{dy}{f(y)} = g(x)dx$$

Cross multiply.

$$\int \frac{1}{f(y)} dy = \int g(x)dx$$

Integrate both sides.

This integral equation implicitly defines y as a function of x. We then solve the equation for y in terms of x, if possible.

Initial value problems consist of a first-order differential equation and an initial condition or value of the equation in the form $y(x_0) = y_0$. To solve initial value problems, we must solve the equation and manipulate the general solution so that the solution satisfies the initial value.

Example:

Solve the initial value problem

$$\frac{dy}{dx} = \frac{2x+1}{2y-2}, \quad y(0) = -1$$

$(2y-2)dy = (2x+1)dx$	Separate variables.
$\int(2y-2)dy = \int(2x+1)dx$	Integrate both sides
$y^2 - 2y = x^2 + x + C$	
$(-1)^2 - 2(-1) = (0)^2 + (0) + C, \quad C = 3$	Apply the initial condition.
$y^2 - 2y = x^2 + x + 3$	Implicit solution.
$y^2 - 2y - (x^2 + x + 3) = 0$	Rewrite equation in quadratic form.
$y(x) = \dfrac{2 \pm \sqrt{(-2)^2 - 4(1)(-1(x^2 + x + 3))}}{2(1)}$	Quadratic formula.
$y(x) = \dfrac{2 \pm \sqrt{4 + 4(x^2 + x + 3)}}{2}$	Simplify.
$y(x) = \dfrac{2 \pm 2\sqrt{1 + (x^2 + x + 3)}}{2} = 1 \pm \sqrt{1 + (x^2 + x + 3)} = 1 \pm \sqrt{x^2 + x + 4}$	

$-1 = y(0) = 1 \pm \sqrt{0^2 + 0 + 4} = 1 \pm 2 = 3 \text{ or } -1$ Apply initial condition.

In this case, the "–" sign produces the correct value, so the final, explicit solution of the initial value problem is

$$y(x) = 1 - \sqrt{x^2 + x + 4}$$

We can use first-order differential equations to model and solve real-world problems. For example, we can solve the following problem involving the flow of water out of a tank using differential equations.

Example:

Let $y(t)$ be the height of the water in a tank at time t. Let $V(t)$ be the volume of water in a tank at time t. Water leaks through a hole of area a out of the bottom of the tank. The following equation defines the rate of change in the volume of water in the tank.

$$\frac{dV}{dt} = -a\sqrt{2gy}$$ where g is the acceleration due to gravity

Suppose the tank is a cylinder with height of 3 feet, a radius of 1 foot, and a circular hole with radius of 2 inches. Find the height of the water at time t, if the tank was full at $t=0$. Assume g equals 32 ft/s^2.

$$V = 3\pi \text{ ft}^3 \qquad a = 4\pi \text{ in}^2 = \frac{1}{36}\pi \text{ ft}^2$$

Define volume of tank and area of hole.

$$\frac{dV}{dt} = -a\sqrt{2gy} = -\frac{1}{36}\pi\sqrt{64y} = -\frac{2}{9}\pi\sqrt{y}$$

Apply rate of change formula.

$$y(t) = \frac{V(t)}{\pi}$$

$$\frac{dy}{dt} = \frac{1}{\pi}\frac{dV}{dt} = (\frac{1}{\pi})(-\frac{2}{9}\pi\sqrt{y}) = -\frac{2}{9}\sqrt{y}$$

Define $y(t)$ and $\frac{dy}{dt}$.

$$\frac{-9}{2\sqrt{y}}dy = dt, \quad \int\frac{-9}{2\sqrt{y}}dy = \int dt$$

Solve and integrate.

$$t = \frac{(-9)2\sqrt{y}}{2} = (-9)\sqrt{y} + C$$

Solve for t.

at $t = 0$, $y = 3$ because the tank is full

$$0 = (-9)\sqrt{3} + C$$
$$C = 9\sqrt{3}$$

Find the constant .

$$t = \frac{(-9)2\sqrt{y}}{2} = (-9)\sqrt{y} + 9\sqrt{3}$$

$$y(t) = \frac{(t-9\sqrt{3})^2}{(-9)^2} = \frac{t^2 - 18\sqrt{3} + 243}{81}$$

Solve for $y(t)$.

SUBAREA V.   PROBABILITY, STATISTICS, AND DISCRETE MATHEMATICS

COMPETENCY 0020   UNDERSTAND THE PRINCIPLES, PROPERTIES, AND TECHNIQUES OF PROBABILITY.

Dependent and Independent Events

Dependent events occur when the probability of the second event depends on the outcome of the first event. For example, consider the two events (A) it is sunny on Saturday and (B) you go to the beach. If you intend to go to the beach on Saturday, rain or shine, then A and B may be independent. If however, you plan to go to the beach only if it is sunny, then A and B may be dependent. In this situation, the probability of event B will change depending on the outcome of event A.

Suppose you have a pair of dice, one red and one green. If you roll a three on the red die and then roll a four on the green die, we can see that these events do not depend on the other. The total probability of the two independent events can be found by multiplying the separate probabilities.

$$P(A \text{ and } B) = P(A) \times P(B)$$
$$= 1/6 \times 1/6$$
$$= 1/36$$

Many times, however, events are not independent. Suppose a jar contains 12 red marbles and 8 blue marbles. If you randomly pick a red marble, replace it and then randomly pick again, the probability of picking a red marble the second time remains the same. However, if you pick a red marble, and then pick again without replacing the first red marble, the second pick becomes dependent upon the first pick.

$$P(\text{Red and Red}) \text{ with replacement} = P(\text{Red}) \times P(\text{Red})$$
$$= 12/20 \times 12/20$$
$$= 9/25$$

$$P(\text{Red and Red}) \text{ without replacement} = P(\text{Red}) \times P(\text{Red})$$
$$= 12/20 \times 11/19$$
$$= 33/95$$

Predicting Odds

Odds are defined as the ratio of the number of favorable outcomes to the number of unfavorable outcomes. The sum of the favorable outcomes and the unfavorable outcomes should always equal the total possible outcomes.

For example, given a bag of 12 red and 7 green marbles compute the odds of randomly selecting a red marble.

$$\text{Odds of red} = \frac{12}{19}$$

$$\text{Odds of not getting red} = \frac{7}{19}$$

In the case of flipping a coin, it is equally likely that a head or a tail will be tossed. The odds of tossing a head are 1:1. This is called even odds.

The Addition Principle of Counting states:

If A and B are events, $n(A \, or \, B) = n(A) + n(B) - n(A \cap B)$.

Example:

In how many ways can you select a black card or a Jack from an ordinary deck of playing cards?

Let B denote the set of black cards and let J denote the set of Jacks. Then,

$$n(B) = 26, n(J) = 4, n(B \cap J) = 2$$
$$= 26 + 4 - 2$$
$$= 28$$

The Addition Principle of Counting for Mutually Exclusive Events states:

If A and B are mutually exclusive events, $n(A \, or \, B) = n(A) + n(B)$.

Example:

A travel agency offers 40 possible trips: 14 to Asia, 16 to Europe and 10 to South America. In how many ways can you select a trip to Asia or Europe through this agency?

Let A denote trips to Asia and let E denote trips to Europe. Then, $A \cap E = \varnothing$ and

$$n(A\,or\,E) = 14 + 16 = 30.$$

Therefore, the number of ways you can select a trip to Asia or Europe is 30.

The Multiplication Principle of Counting for Dependent Events states:

Let A be a set of outcomes of Stage 1 and B a set of outcomes of Stage 2. Then the number of ways $n(A\,and\,B)$, that A and B can occur in a two-stage experiment is given by:

$$n(A\,and\,B) = n(A)n(B\,|\,A),$$

where $n(B\,|\,A)$ denotes the number of ways B can occur given that A has already occurred.

Example:

How many ways from an ordinary deck of 52 cards can 2 Jacks be drawn in succession if the first card is drawn but not replaced in the deck and then the second card is drawn?

This is a two-stage experiment for which we wish to compute $n(A\,and\,B)$, where A is the set of outcomes for which a Jack is obtained on the first draw and B is the set of outcomes for which a Jack is obtained on the second draw.

If the first card drawn is a Jack, then there are only three remaining Jacks left to choose from on the second draw. Thus, drawing two cards without replacement means the events A and B are dependent.

$$n(A\,and\,B) = n(A)n(B\,|\,A) = 4 \cdot 3 = 12$$

The Multiplication Principle of Counting for Independent Events states:

Let A be a set of outcomes of Stage 1 and B a set of outcomes of Stage 2. If A and B are independent events, then the number of ways $n(A\,and\,B)$, that A and B can occur in a two-stage experiment is given by:

$$n(A\,and\,B) = n(A)n(B).$$

Example:

How many six-letter code "words" can be formed if repetition of letters is not allowed?

Since these are code words, a word does not have to look like a word; for example, abcdef could be a code word. Since we must choose a first letter *and* a second letter *and* a third letter *and* a fourth letter *and* a fifth letter *and* a sixth letter, this experiment has six stages.

Since repetition is not allowed there are 26 choices for the first letter; 25 for the second; 24 for the third; 23 for the fourth; 22 for the fifth; and 21 for the sixth. Therefore, we have:

n(six-letter code words without repetition of letters)

$$= 26 \cdot 25 \cdot 24 \cdot 23 \cdot 22 \cdot 21$$

$$= 165,765,600$$

Histograms are used to summarize information from large sets of data that can be naturally grouped into intervals. The vertical axis indicates **frequency** (the number of times any particular data value occurs), and the horizontal axis indicates data values or ranges of data values. The number of data values in any interval is the **frequency of the interval**.

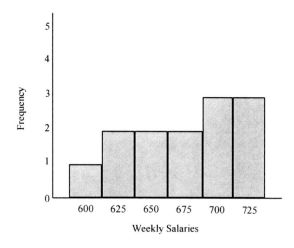

Bar graphs are similar to histograms. However, bar graphs are often used to convey information about categorical data where the horizontal scale represents a non-numeric attributes such as cities or years. Another difference is that the bars in bar graphs rarely touch. Bar graphs are also useful in comparing data about two or more similar groups of items.

**Production for ACME
June-September 2006
(in thousands)**

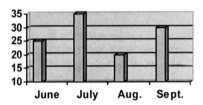

A **pie chart**, also known as a **circle graph**, is used to represent relative amounts of a whole.

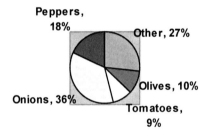

Scatter plots compare two characteristics of the same group of things or people and usually consist of a large body of data. They show how much one variable is affected by another. The relationship between the two variables is their **correlation**. The closer the data points come to making a straight line when plotted, the closer the correlation.

Stem and leaf plots are visually similar to line plots. The **stems** are the digits in the greatest place value of the data values, and the **leaves** are the digits in the next greatest place values. Stem and leaf plots are best suited for small sets of data and are especially useful for comparing two sets of data. The following is an example using test scores:

4	9
5	4 9
6	1 2 3 4 6 7 8 8
7	0 3 4 6 6 6 7 7 7 8 8 8 8
8	3 5 5 7 8
9	0 0 3 4 5
10	0 0

A **pictograph** uses small figures or icons to represent data. Pictographs are used to summarize relative amounts, trends, and data sets. They are useful in comparing quantities.

Monarch Butterfly Migration to the U.S.
in millions

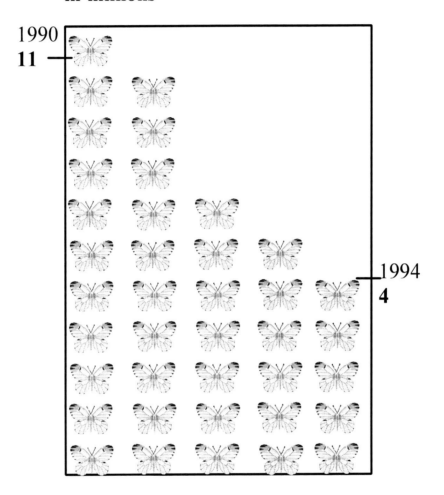

The data in this graph is not accurate. It is for illustration purposes only.

The **binomial distribution** is a sequence of probabilities with each probability corresponding to the likelihood of a particular event occurring. It is called a binomial distribution because each trial has precisely two possible outcomes. An **event** is defined as a sequence of Bernoulli trials that has within it a specific number of successes. The order of success is not important.

Note: There are two parameters to consider in a binomial distribution:

1. p = the probability of a success
2. n = the number of Bernoulli trials (i.e., the length of the sequence).

Example:

Toss a coin two times. Each toss is a Bernoulli trial as discussed above. Consider heads to be success. One event is one sequence of two coin tosses. Order does not matter.

There are two possibilities for each coin toss. Therefore, there are four (2·2) possible subevents: 00, 01, 10, 11 (where 0 = tail and 1 = head).

According to the multiplication rule, each subevent has a probability of $\frac{1}{4}\left(\frac{1}{2}\cdot\frac{1}{2}\right)$.

One subevent has zero heads, so the event of zero heads in two tosses is $p(h=0)=\frac{1}{4}$.

Two subevents have one head, so the event of one head in two tosses is $p(h=1)=\frac{2}{4}$.

One subevent has two heads, so the event of two heads in two tosses is $p(h=2)=\frac{1}{4}$.

So the binomial distribution for two tosses of a fair coin is:

$$p(h=0)=\frac{1}{4}, \quad p(h=1)=\frac{2}{4}, \quad p(h=2)=\frac{1}{4}.$$

A **normal distribution** is the distribution associated with most sets of real-world data. It is frequently called a **bell curve**. A normal distribution has a **random variable** X with mean μ and variance σ^2.

Example:

Albert's Bagel Shop's morning customer load follows a normal distribution, with **mean** (average) 50 and **standard deviation** 10. The standard deviation is the measure of the variation in the distribution. Determine the probability that the number of customers tomorrow will be less than 42.

First convert the raw score to a **z-score**. A z-score is a measure of the distance in standard deviations of a sample from the mean.

The z-score = $\dfrac{X_i - \bar{X}}{s} = \dfrac{42 - 50}{10} = \dfrac{-8}{10} = -.8$

Next, use a table to find the probability corresponding to the z-score. The table gives us .2881. Since our raw score is negative, we subtract the table value from .5.

$$.5 - .2881 = .2119$$

We can conclude that $P(x < 42) = .2119$. This means that there is about a 21% chance that there will be fewer than 42 customers tomorrow morning.

Example:

The scores on Mr. Rogers' statistics exam follow a normal distribution with mean 85 and standard deviation 5. A student is wondering what the probability is that she will score between a 90 and a 95 on her exam.

We wish to compute $P(90 < x < 95)$.

Compute the z-scores for each raw score.

$$\frac{90 - 85}{5} = \frac{5}{5} = 1 \text{ and } \frac{95 - 85}{5} = \frac{10}{5} = 2 .$$

Now we want $P(1 < z < 2)$.

Since we are looking for an occurrence between two values, we subtract:

$P(1 < z < 2) = P(z < 2) - P(z < 1).$

We use a table to get:
$P(1 < z < 2) = .9772 - .8413 = .1359.$

(Remember that since the z-scores are positive, we add .5 to each probability.)

We can then conclude that there is a 13.6% chance that the student will score between a 90 and a 95 on her exam.

COMPETENCY 0021 **UNDERSTAND THE PRINCIPLES, PROPERTIES, AND TECHNIQUES OF STATISTICS.**

Random sampling supplies every combination of items from the frame, or stratum, as a known probability of occurring. A large body of statistical theory quantifies the risk and thus enables an appropriate sample size to be chosen.

Systematic sampling selects items in the frame according to the sample. The first item is chosen to be the , where is a random integer in the range .

There are three stages to Cluster or Area sampling: the target population is divided into many regional clusters (groups); a few clusters are randomly selected for study; a few subjects are randomly chosen from within a cluster.

Convenience sampling is the method of choosing items arbitrarily and in an unstructured manner from the frame.

To make a **bar graph** or a **pictograph**, determine the scale to be used for the graph. Then determine the length of each bar on the graph or determine the number of pictures needed to represent each item of information. Be sure to include an explanation of the scale in the legend.

Example: A class had the following grades:
    4 A's, 9 B's, 8 C's, 1 D, 3 F's.
    Graph these on a bar graph and a pictograph.

Pictograph

Grade	Number of Students
A	😊😊😊😊
B	😊😊😊😊😊😊😊😊😊
C	😊😊😊😊😊😊😊😊
D	😊
F	😊😊😊

Bar raph

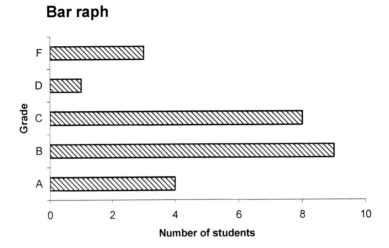

To make a **line graph**, determine appropriate scales for both the vertical and horizontal axes (based on the information to be graphed). Describe what each axis represents and mark the scale periodically on each axis. Graph the individual points of the graph and connect the points on the graph from left to right.

Example: Graph the following information using a line graph.

The number of National Merit finalists/school year

	90-'91	91-'92	92-'93	93-'94	94-'95	95-'96
Central	3	5	1	4	6	8
Wilson	4	2	3	2	3	2

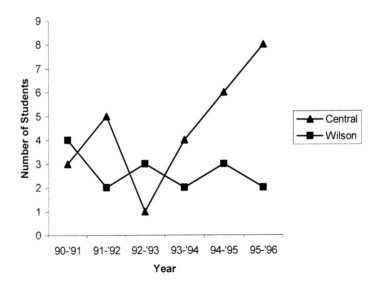

To make a **circle graph**, total all the information that is to be included on the graph. Determine the central angle to be used for each sector of the graph using the following formula:

$$\frac{\text{information}}{\text{total information}} \times 360° = \text{degrees in central } \square$$

Lay out the central angles to these sizes, label each section and include its percent.

Example: Graph this information on a circle graph:

Monthly expenses:

Rent, $400
Food, $150
Utilities, $75
Clothes, $75
Church, $100
Misc., $200

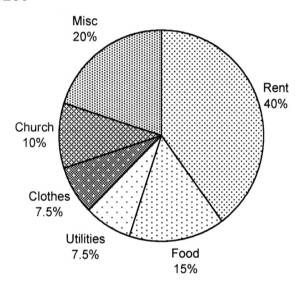

To read a bar graph or a pictograph, read the explanation of the scale that was used in the legend. Compare the length of each bar with the dimensions on the axes and calculate the value each bar represents. On a pictograph count the number of pictures used in the chart and calculate the value of all the pictures.

To read a circle graph, find the total of the amounts represented on the entire circle graph. To determine the actual amount that each sector of the graph represents, multiply the percent in a sector times the total amount number.

To read a chart read the row and column headings on the table. Use this information to evaluate the given information in the chart.

Percentiles, Stanines, and Quartiles – Graphical Representations

Percentiles divide data into 100 equal parts. A person whose score falls in the 65th percentile has outperformed 65 percent of all those who took the test. This does not mean that the score was 65 percent out of 100 nor does it mean that 65 percent of the questions answered were correct. It means that the grade was higher than 65 percent of all those who took the test.

Stanine "standard nine" scores combine the understandability of percentages with the properties of the normal curve of probability. Stanines divide the bell curve into nine sections, the largest of which stretches from the 40th to the 60th percentile and is the "Fifth Stanine" (the average of taking into account error possibilities).

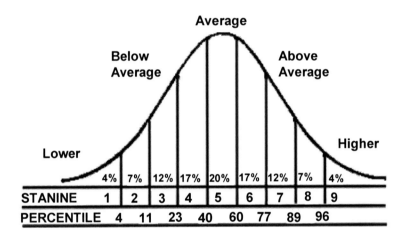

4%	7%	12%	17%	20%	17%	12%	7%	4%
STANINE 1	2	3	4	5	6	7	8	9
PERCENTILE 4	11	23	40	60	77	89	96	

Quartiles divide the data into 4 parts. First find the median of the data set (Q2), then find the median of the upper (Q3) and lower (Q1) halves of the data set. If there are an odd number of values in the data set, include the median value in both halves when finding quartile values. For example, given the data set: {1, 4, 9, 16, 25, 36, 49, 64, 81} first find the median value, which is 25 (this is the second quartile). Since there are an odd number of values in the data set (9), we include the median in both halves. To find the quartile values, we much find the medians of: {1, 4, 9, 16, 25} and {25, 36, 49, 64, 81}. Since each of these subsets had an odd number of elements (5), we use the middle value. Thus the first quartile value is 9 and the third quartile value is 49. If the data set had an even number of elements, average the middle two values. The quartile values are always either one of the data points, or exactly half way between two data points.

Sample problem:

1. Given the following set of data, find the percentile of the score 104. [70, 72, 82, 83, 84, 87, 100, 104, 108, 109, 110, 115]

    Solution: Find the percentage of scores below 104.

    7/12 of the scores are less than 104. This is 58.333%; therefore, the score of 104 is in the 58th percentile.

2. Find the first, second and third quartile for the data listed.
    6, 7, 8, 9, 10, 12, 13, 14, 15, 16, 18, 23, 24, 25, 27, 29, 30, 33, 34, 37

    Quartile 1:   The 1st Quartile is the median of the lower half of the data set, which is 11.

    Quartile 2:   The median of the data set is the 2nd Quartile, which is 17.

    Quartile 3:   The 3rd Quartile is the median of the upper half of the data set, which is 28.

Mean, median, and mode are three measures of central tendency. The **mean** is the average of the data items. The **median** is found by putting the data items in order from smallest to largest and selecting the item in the middle (or the average of the two items in the middle). The **mode** is the most frequently occurring item.

Range is a measure of variability. It is found by subtracting the smallest value from the largest value.

Sample problem:

Find the mean, median, mode, and range of the test scores listed below:

85	77	65
92	90	54
88	85	70
75	80	69
85	88	60
72	74	95

Mean (X)   = sum of all scores ÷ number of scores
                   = $1404 \div 18$
                   = 78

Median = put numbers in order from smallest to largest. Pick middle number.

54, 60, 65, 69, 70, 72, 74, 75, 77, 80, 85, 85, 85, 88, 88, 90, 92, 95

The numbers 77 and 80 are both in the middle. Therefore, the median is the average of the 2 numbers in the middle or 78.5.

Mode = most frequent number
        = 85

Range = largest number minus the smallest number
          = 95 − 54
          = 41

Different situations require different information. If we examine the circumstances under which an ice cream storeowner may use statistics collected in the store, we find different uses for different information.

Over a 7-day period, the store owner collected data on the ice cream flavors sold. He found the mean number of scoops sold was 174 per day. The most frequently sold flavor was vanilla. This information was useful in determining how much ice cream to order in all and in what amounts for each flavor.

In the case of the ice cream store, the median and range had little business value for the owner.

Consider the set of test scores from a math class: 0, 16, 19, 65, 65, 65, 68, 69, 70, 72, 73, 73, 75, 78, 80, 85, 88, and 92. The mean is 64.06 and the median is 71. Since there are only three scores less than the mean out of the eighteen scores, the median (71) would be a more descriptive score.

Retail storeowners may be most concerned with the most common dress size so they may order more of that size than any other size.

An understanding of the definitions is important in determining the validity and uses of statistical data. All definitions and applications in this section apply to ungrouped data.

Data item: each piece of data is represented by the letter X.

Mean: the average of all data represented by the symbol X.

Range: difference between the highest and lowest value of data items.

Sum of the Squares: sum of the squares of the differences between each item and the mean.

$$Sx^2 = (X - X)^2$$

Variance: the sum of the squares quantity divided by the number of items.

(the lower case Greek letter sigma (σ) squared represents variance).

$$\frac{Sx^2}{N} = \sigma^2$$

The larger the value of the variance the larger the spread

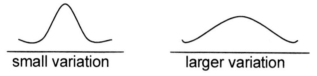

small variation   larger variation

Standard Deviation: the square root of the variance. The lower case Greek letter sigma (σ) is used to represent standard deviation.

$$\sigma = \sqrt{\sigma^2}$$

Most statistical calculators have standard deviation keys on them and should be used when asked to calculate statistical functions. It is important to become familiar with the calculator and the location of the keys needed.

Sample Problem:

Given the ungrouped data below, calculate the mean, range, standard deviation and the variance.

| 15 | 22 | 28 | 25 | 34 | 38 |
| 18 | 25 | 30 | 33 | 19 | 23 |

Mean (X) = 25.8333333
Range: $38 - 15 = 23$
standard deviation (σ) = 6.699137
Variance (σ^2) = 48.87879

Correlation is a measure of association between two variables. It varies from −1 to 1, with 0 being a random relationship, 1 being a perfect positive linear relationship, and −1 being a perfect negative linear relationship.

The **correlation coefficient** (r) is used to describe the strength of the association between the variables and the direction of the association.

Example:

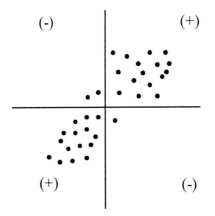

Horizontal and vertical lines are drawn through the point of averages which is the point on the averages of the x and y values. This divides the scatter plot into four quadrants. If a point is in the lower left quadrant, the product of two negatives is positive; in the upper right, the product of two positives is positive. The positive quadrants are depicted with the positive sign (+). In the two remaining quadrants (upper left and lower right), the product of a negative and a positive is negative. The negative quadrants are depicted with the negative sign (−). If r is positive, then there are more points in the positive quadrants and if r is negative, then there are more points in the two negative quadrants.

Regression is a form of statistical analysis used to predict a dependent variable (y) from values of an independent variable (x). A regression equation is derived from a known set of data.

The simplest regression analysis models the relationship between two variables using the following equation: $y = a + bx$, where y is the dependent variable and x is the independent variable. This simple equation denotes a linear relationship between x and y. This form would be appropriate if, when you plotted a graph of x and y, you tended to see the points roughly form along a straight line.

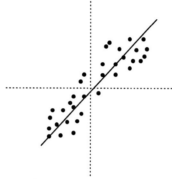

The line can then be used to make predictions.

If all of the data points fell on the line, there would be a perfect correlation ($r = 1.0$) between the x and y data points. These cases represent the best scenarios for prediction. A positive or negative r value represents how y varies with x. When r is positive, y increases as x increases. When r is negative y decreases as x increases.

COMPETENCY 0022   UNDERSTAND THE PRINCIPLES OF DISCRETE MATHEMATICS.

The difference between **permutations and combinations** is that in permutations all possible ways of writing an arrangement of objects are given, while in a combination, a given arrangement of objects is listed only once.

Given the set {1, 2, 3, 4}, list the arrangements of two numbers that can be written as a combination and as a permutation.

Combination	Permutation
12, 13, 14, 23, 24, 34	12, 21, 13, 31, 14, 41, 23, 32, 24, 42, 34, 43,
six ways	twelve ways

Using the formulas given below the same results can be found.

$$_nP_r = \frac{n!}{(n-r)!}$$

The notation $_nP_r$ is read "the number of permutations of n objects taken r at a time."

$$_4P_2 = \frac{4!}{(4-2)!}$$

Substitute known values.

$$_4P_2 = 12$$

Solve.

$$_nC_r = \frac{n!}{(n-r)!r!}$$

The number of combinations when r objects are selected from n objects.

$$_4C_2 = \frac{4!}{(4-2)!2!}$$

Substitute known values.

$$_4C_2 = 6$$

Solve.

Possibly the most famous sequence is the **Fibonacci sequence**. A basic Fibonacci sequence is when two numbers are added together to get the next number in the sequence. An example would be 1, 1, 2, 3, 5, 8, 13, ….

Arithmetic Sequences

When given a set of numbers where the common difference between the terms is constant, use the following formula:

$a_n = a_1 + (n-1)d$, where

> a_1 = the first term
> a_n = the nth term (general term)
> d = the common difference

Example:
Find the 8th term of the arithmetic sequence 5, 8, 11, 14...

$a_n = a_1 + (n-1)d$	
$a_1 = 5$	identify the 1st term
$d = 8 - 5 = 3$	find d
$a_n = 5 + (8-1)3$	substitute
$a_n = 26$	

Example:
Given two terms of an arithmetic sequence, find a_1 and d.

$a_4 = 21$	$a_6 = 32$
$a_n = a_1 + (n-1)d$	$a_4 = 21, n = 4$
$21 = a_1 + (4-1)d$	$a_6 = 32, n = 6$
$32 = a_1 + (6-1)d$	

$21 = a_1 + 3d$	solve the system of equations
$32 = a_1 + 5d$	

$32 = a_1 + 5d$	
$\underline{-21 = -a_1 - 3d}$	multiply by -1
$11 = 2d$	add the equations

$5.5 = d$

$21 = a_1 + 3(5.5)$	substitute d = 5.5 into one of the equations
$21 = a_1 + 16.5$	
$a_1 = 4.5$	

The sequence begins with 4.5 and has a common difference of 5.5 between numbers.

Geometric Sequences

When using geometric sequences, consecutive numbers are compared to find the common ratio.

$$r = \frac{a_{n+1}}{a_n} \text{ where}$$

r = common ratio

a_n = the nth term

The ratio is then used in the geometric sequence formula:
$a_n = a_1 r^{n-1}$

<u>Example:</u>
Find the 8th term of the geometric sequence 2, 8, 32, 128...

$$r = \frac{a_{n+1}}{a_n} \qquad \text{use common ratio formula to find ratio}$$

$$r = \frac{8}{2} \qquad \text{substitute } a_n = 2, \; a_{n+1} = 8$$

$$r = 4$$

$$a_n = a_1 \bullet r^{n-1} \qquad \text{use } r = 4 \text{ to solve for the 8th term}$$
$$a_n = 2 \bullet 4^{8-1}$$
$$a_n = 32,768$$

Summation of Terms

The sums of terms in a progression is simply found by determining if it is an arithmetic or geometric sequence and then using the appropriate formula.

Sum of first n terms of

an arithmetic sequence.

$$S_n = \frac{n}{2}(a_1 + a_n)$$

or

$$S_n = \frac{n}{2}\left[2a_1 + (n-1)d\right]$$

Sum of first n terms of

a geometric sequence.

$$S_n = \frac{a_1\left(r^n - 1\right)}{r-1}, r \neq 1$$

Sample Problems:

1. $\displaystyle\sum_{i=1}^{10}(2i+2)$

This means find the sum of the terms beginning with the first term and ending with the 10th term of the sequence $a = 2i + 2$.

$a_1 = 2(1) + 2 = 4$

$a_{10} = 2(10) + 2 = 22$

$S_n = \dfrac{n}{2}(a_1 + a_n)$

$S_n = \dfrac{10}{2}(4 + 22)$

$S_n = 130$

2. Find the sum of the first 6 terms in an arithmetic sequence if the first term is 2 and the common difference, d is-3.

$n = 6 \qquad a_1 = 2 \qquad d = {}^-3$

$S_n = \dfrac{n}{2}\big[2a_1 + (n-1)d\big]$

$S_6 = \dfrac{6}{2}\big[2\times2 + (6-1)\,{}^-3\big]$ Substitute known values.

$S_6 = 3\big[4 + ({}^-15)\big]$ Solve.

$S_6 = 3(-11) = -33$

3. Find $\displaystyle\sum_{i=1}^{5}4\times2^{\,i}$

This means the sum of the first 5 terms where $a_i = a\times b^i$ and $r = b$.

$a_1 = 4\times2^1 = 8$ Identify a_1, r, n

$r = 2 \qquad n = 5$

$S_n = \dfrac{a_1(r^n - 1)}{r - 1}$ Substitute a, r, n

$S_5 = \dfrac{8(2^5 - 1)}{2 - 1}$ Solve.

$S_5 = \dfrac{8(31)}{1} = 248$

Practice problems:

1. Find the sum of the first five terms of the sequence if $a = 7$ and $d = 4$.

2. $\displaystyle\sum_{i=1}^{7} (2i - 4)$

3. $\displaystyle\sum_{i=1}^{6} {}^{-}3\left(\frac{2}{5}\right)^{i}$

Sequences can be **finite** or **infinite**. A finite sequence is a sequence whose domain consists of the set {1, 2, 3, ... *n*} or the first *n* positive integers. An infinite sequence is a sequence whose domain consists of the set {1, 2, 3, ...}; which is in other words all positive integers.

A **recurrence relation** is an equation that defines a sequence recursively; in other words, each term of the sequence is defined as a function of the preceding terms.

A real-life application would be using a recurrence relation to determine how much your savings would be in an account at the end of a certain period of time. For example:

You deposit $5,000 in your savings account. Your bank pays 5% interest compounded annually. How much will your account be worth at the end of 10 years?

Let V represent the amount of money in the account and V_n represent the amount of money after n years.

The amount in the account after n years equals the amount in the account after $n - 1$ years plus the interest for the nth year. This can be expressed as the recurrence relation V_0 where your initial deposit is represented by $V_0 = 5,000$.

$$V_0 = V_0$$
$$V_1 = 1.05V_0$$
$$V_2 = 1.05V_1 = (1.05)^2 V_0$$
$$V_3 = 1.05V_2 = (1.05)^3 V_0$$
$$\ldots\ldots$$
$$V_n = (1.05)V_{n-1} = (1.05)^n V_0$$

Inserting the values into the equation, you get
$V_{10} = (1.05)^{10}(5,000) = 8,144$.

You determine that after investing $5,000 in an account earning 5% interest, compounded annually for 10 years, you would have $8,144.

CURRICULUM AND INSTRUCTION

The National Council of Teachers of Mathematics standards emphasize the teacher's obligation to make mathematics relevant to the students and applicable to the real world. The mathematics need in our technological society is different from that need in the past; we need thinking skills rather than computational. Mathematics needs to connect to other subjects, as well as other areas of math.

ERROR ANALYSIS

A simple method for analyzing student errors is to ask how the answer was obtained. The teacher can then determine if a common error pattern has resulted in the wrong answer. There is a value to having the students explain how the arrived at the correct as well as the incorrect answers.

Many errors are due to simple **carelessness**. Students need to be encouraged to work slowly and carefully. They should check their calculations by redoing the problem on another paper, not merely looking at the work. Addition and subtraction problems need to be written neatly so the numbers line up. Students need to be careful regrouping in subtraction. Students must write clearly and legibly, including erasing fully. Use estimation to ensure that answers make sense.

Many students' computational skills exceed their **reading** level. Although they can understand basic operations, they fail to grasp the concept or completely understand the question. Students must read directions slowly.

Fractions are often a source of many errors. Students need to be reminded to use common denominators when adding and subtracting and to always express answers in simplest terms. Again, it is helpful to check by estimating.

The most common error that is made when working with **decimals** is failure to line up the decimal points when adding or subtracting or not moving the decimal point when multiplying or dividing. Students also need to be reminded to add zeroes when necessary. Reading aloud may also be beneficial. Estimation, as always, is especially important.

Students need to know that it is okay to make mistakes. The teacher must keep a positive attitude, so they do not feel defeated or frustrated.

REPRESENTATIONS OF CONCEPTS

Mathematical operations can be shown using manipulatives or drawings. Multiplication can be shown using arrays.

3×4

Addition and subtractions can be demonstrated with symbols.

$\psi \, \psi \, \psi \, \xi \, \xi \, \xi \, \xi$

$3 + 4 = 7$

$7 - 3 = 4$

Fractions can be clarifies using pattern blocks, fraction bars, or paper folding.

CONCEPT DEVELOPMENT

Manipulatives can foster learning for all students. Mathematics needs to be derived from something that is real to the learner. If he can "touch" it, he will understand and remember. Students can use fingers, ice cream sticks, tiles and paper folding, as well as those commercially available manipulatives to visualize operations and concepts. The teacher needs to solidify the concrete examples into abstract mathematics.

PROBLEM SOLVING

Problem solving strategies are simply plans of attack. Student often panic when confronted with word problems. If they have a "list" of ideas, ways to attempt a solution, they will be able to approach the problems more calmly and confidently. Some methods include, but are not limited to, draw a diagram, work backwards, guess and check, and solve a simpler problem.

It is helpful to have students work in groups. Mathematics does not have to be solitary activity. Cooperative learning fosters enthusiasm. Creating their own problems is another useful tool. Also, encourage students to find more than one way to solve a problem. Thinking about problem solving after the solution has been discovered encourages understanding and creativity. The more they practice problems, the more comfortable and positive students will feel.

MATHEMATICAL LANGUAGE

Students need to use the proper mathematical terms and expressions. When reading decimals, they need to read 0.4 as "four tenths" to promote better understanding of the concepts. They should do their work in a neat and organized manner. Students need to be encouraged to verbalize their strategies, both in computation and word problems. Additionally, writing original word problems fosters understanding of math language. Another idea is requiring students to develop their own mathematical glossary. Knowing the answers and being able to communicate them are equally important.

MANIPULATIVES

<u>Example</u>:
Using tiles to demonstrate both geometric ideas and number theory.

Give each group of students 12 tiles and instruct them to build rectangles. Students draw their rectangles on paper.

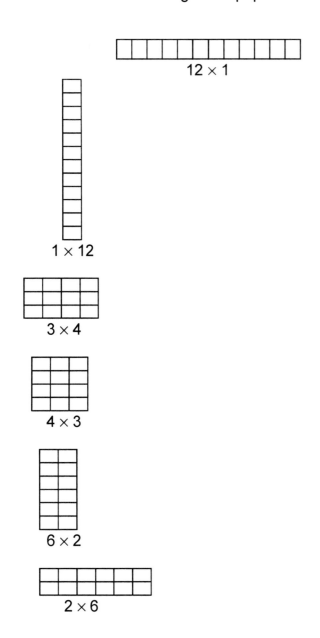

Encourage students to describe their reactions. Extend to 16 tiles. Ask students to form additional problems.

CALCULATORS

Calculators are an important tool. They should be encouraged in the classroom and at home. They do not replace basic knowledge but they can relieve the tedium of mathematical computations, allowing students to explore more challenging mathematical directions. Students will be able to use calculators more intelligently if they are taught how. Students need to always check their work by estimating. The goal of mathematics is to prepare the child to survive in the real world. Technology is a reality in today's society.

CHILD DEVELOPMENT

Means of instruction need to be varied to reach children of different skill levels and learning styles. In addition to directed instruction, students should work cooperatively, explore hands-on activities and do projects.

COMPUTERS

Computers can not replace teachers. However, they can be used to enhance the curriculum. They may be used cautiously to help students practice basic skills. Many excellent programs exist to encourage higher-order thinking skills, creativity and problem solving. Learning to use technology appropriately is an important preparation for adulthood. Computers can also show the connections between mathematics and the real world.

QUESTIONING TECHNIQUES

As the teacher's role in the classroom changes from lecturer to facilitator, the questions need to further stimulate students in various ways.

- Helping students work together

What do you think about what John said?

Do you agree? Disagree?

Can anyone explain that differently?

- Helping students determine for themselves if an answer is correct

What do you think that is true?

How did you get that answer?

Do you think that is reasonable? Why?

- Helping students learn to reason mathematically

Will that method always work?

Can you think of a case where it is not true?

How can you prove that?

Is that answer true in all cases?

- Helping student brainstorm and problem solve

Is there a pattern?

What else can you do?

Can you predict the answer?

What if...?

- Helping students connect mathematical ideas

What did we learn before that is like this?

Can you give an example?

What math did you see on television last night? in the newspaper?

Answer Key to Practice Problems

Competency 0002, pg. 8

Question #1   The Red Sox won the World Series.
Question #2   Angle B is not between 0 and 90 degrees.
Question #3   Annie will do well in college.
Question #4   You are witty and charming.

Competency 0003, pg. 23

Question #1 $\dfrac{^{-}12x^5z^5}{y}$

Question #2 $25a^4 + 15a^2 - 7b^3$

Question #3 $100x^{20}y^8 + 8x^9y^6$

Competency 0005, pg. 34

Question #2 a, b, c, f are functions
Question #3 Domain = $^{-}\infty, \infty$ Range = $^{-}5, \infty$

pg. 41

Question #1 Domain = $^{-}\infty, \infty$ Range = $^{-}6, \infty$
Question #2 Domain = 1,4,7,6   Range = -2
Question #3 Domain = $x \neq 2, ^{-}2$
Question #4 Domain = $^{-}\infty, \infty$ Range = -4, 4
                      Domain = $^{-}\infty, \infty$ Range = 2, ∞
Question #5 Domain = $^{-}\infty, \infty$ Range = 5
Question #6 (3,9), (-4,16), (6,3), (1,9), (1,3)

Competency 0006, pg. 44

Question #1 $\begin{pmatrix} 11 & 7 \\ 11 & 1 \end{pmatrix}$

Question #2 $\begin{pmatrix} 0 & 3 \\ -10 & 13 \\ 5 & 0 \end{pmatrix}$

pg. 45

Question #1 $\begin{pmatrix} -4 & 0 & -2 \\ 2 & 4 & -8 \end{pmatrix}$

Question #2 $\begin{pmatrix} 18 \\ 34 \\ 32 \end{pmatrix}$

Question #3 $\begin{pmatrix} -12 & 16 \\ -4 & -2 \\ 0 & 6 \end{pmatrix}$

pg. 47

Question #1 $\begin{pmatrix} -15 & 25 \\ -1 & -13 \end{pmatrix}$

Question #2 $\begin{pmatrix} 5 & -5 & -10 \\ 5 & 5 & 0 \\ 1 & 8 & 7 \\ -9 & 13 & 22 \end{pmatrix}$

pg. 49

Question #1 $\begin{pmatrix} x \\ y \end{pmatrix} = \begin{pmatrix} 3 \\ 1 \end{pmatrix}$

Question #2 $\begin{pmatrix} x \\ y \\ z \end{pmatrix} = \begin{pmatrix} 4 \\ 4 \\ 1 \end{pmatrix}$

pg. 50

Question #1 $\begin{pmatrix} -7 & 0 \\ -12 & -3 \\ 6 & 5 \end{pmatrix}$

Question #2 $\begin{pmatrix} 3 & \dfrac{47}{4} \\ 3 & 7 \end{pmatrix}$

Question #3 $a = -5$ $b = -1$ $c = 0$ $d = -4$ $e = 0$ $f = -10$

Competency 0007, pg. 52

Question #1   x-intercept = -14   y-intercept = -10   slope = $-\dfrac{5}{7}$

Question #2   x-intercept = 14   y-intercept = -7   slope = $\dfrac{1}{2}$

Question #3   x-intercept = 3   y-intercept = none

Question #4   x-intercept = $\dfrac{15}{2}$ y-intercept = 3   slope = $-\dfrac{2}{5}$

pg. 53

Question #1 $y = \dfrac{3}{4}x + \dfrac{17}{4}$

Question #2 $x = 11$

Question #3 $y = \dfrac{3}{5}x + \dfrac{42}{5}$

Question #4 $y = 5$

pg. 55

Question #1   Question #2

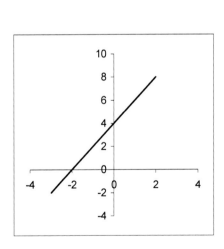   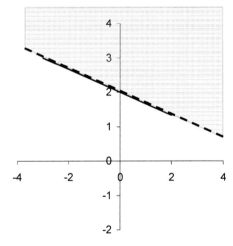

Question #3

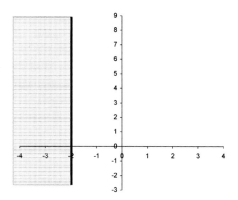

Competency 0008, pg. 75

Question #1 $(6x - 5y)(36x^2 + 30xy + 25y^2)$

Question #2 $4(a - 2b)(a^2 + 2ab + 4b^2)$

Question #3 $5x^2(2x^9 + 3y)(4x^{18} - 6x^9y + 9y^2)$

pg.80

Question #1   The sides are 8, 15, and 17

Question #2   The numbers are 2 and $\dfrac{1}{2}$

pg. 87

Question #1 $x = \dfrac{7 \pm \sqrt{241}}{12}$

Question #2 $x = \dfrac{1}{2}$ or $\dfrac{^-2}{5}$

Question #3 $x = 2$ or $\dfrac{6}{5}$

pg. 88

Question #1 discriminant = 241; 2 real irrational roots
Question #2 discriminant = 81; 2 real rational roots
Question #3 discriminant = 0; 1 real rational root

Competency 0009, pg. 99

Question #1 $\dfrac{14x + 28}{(x+6)(x+1)(x-1)}$

Question #2 $\dfrac{x^3 - 5x^2 + 10x - 12}{x^2 + 3x - 10}$

pg. 101

Question #1 $\dfrac{8x+36}{(x-3)(x+7)}$

Question #2 $\dfrac{25a^2+12b^2}{20a^4b^5}$

Question #3 $\dfrac{2x^2+5x-21}{(x-5)(x+5)(x+3)}$

pg. 102

Question #1 $9xz^5$

Question #2 $\dfrac{3x+2y}{x^2+5xy+25y^2}$

Question #3 $\dfrac{x^2+8x+15}{(x+2)(x+3)(x-7)}$

pg. 104

Question #1 $17\sqrt{6}$

Question #2 $84x^4y^8\sqrt{2}$

Question #3 $\dfrac{5a^4\sqrt{35b}}{8b}$

Question #4 $-5\left(3+\sqrt{3}+2\sqrt{5}\right)$

pg. 105

Question #1 $6a^4\sqrt{2a}$

Question #2 $7i\sqrt{2}$

Question #3 $-2x^2$

Question #4 $6x^4y^3\sqrt[4]{3x^2y^3}$

pg. 106

Question #1 $5\sqrt{6}$

Question #2 $-5\sqrt{3}+6\sqrt{5}=\sqrt{15}-30$

Question #3 $-3\sqrt{2}-9\sqrt{3}-3\sqrt{6}-12$

pg. 110

Question #1

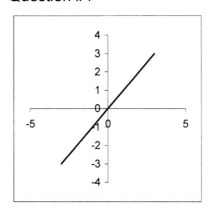

Question #2

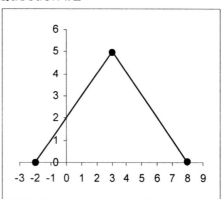

Question #3

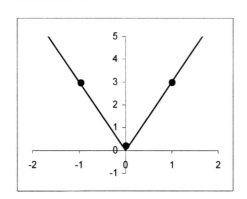

Questions#4

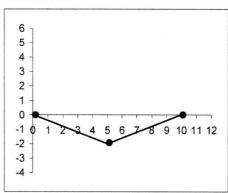

pg. 112

Question #1 It takes Curly 15 minutes to paint the elephant alone
Question #2 The original number is 5/15.
Question #3 The car was traveling at 68mph and the truck was traveling at 62mph

pg. 113

Question #1 $C = \dfrac{5}{9}F - \dfrac{160}{9}$

Question #2 $b = \dfrac{2A - 2h^2}{h}$

Question #3 $n = \dfrac{360 + S}{180}$

pg. 114

Question #1 $x = 7,\ x = ^- 5$

Question #2 $x = \dfrac{13}{8}$

pg. 116

    Question #1 $x = 11$
    Question #2 $x = 17$

Competency 0010, pg. 124

    Question #2 $x = 9$
    Question #3 $x = 3$ or $x = 6$
    Question #4   1

Competency 0017, pg. 193

    Question #1 $\cot\theta = \dfrac{x}{y}$

$$\frac{x}{y} = \frac{x}{r} \times \frac{r}{y} = \frac{x}{y} = \cot\theta$$

    Question #2 $1 + \cot^2\theta = \csc^2\theta$

$$\frac{y^2}{y^2} + \frac{x^2}{y^2} = \frac{r^2}{y^2} = \csc^2\theta$$

Competency 0018, pg. 201

    Question #1   49.34
    Question #2   1

pg. 203

    Question #1 ∞
    Question #2   -1

Competency 0019, pg. 229

    Question #1 $g(x) = -\sin x + c$
    Question #2 $g(x) = \pi \sec x + c$

pg. 235

    Question #1 $t(0) = -24$ m/sec
    Question #2 $t(4) = 24$ m/sec

Competency 0022, pg. 265

    Question #1 $S_5 = 75$
    Question #2 $S_n = 28$
    Question #3 $S_n = -\dfrac{-31122}{15625} \approx {}^- 1.99$

Sample Test

1. Change $.\overline{63}$ into a fraction in simplest form.

   A) $63/100$
   B) $7/11$
   C) $6 \ 3/10$
   D) $2/3$

2. Which of the following sets is closed under division?

   I) $\{½, 1, 2, 4\}$
   II) $\{-1, 1\}$
   III) $\{-1, 0, 1\}$

   A) I only
   B) II only
   C) III only
   D) I and II

3. Which of the following illustrates an inverse property?

   A) a + b = a - b
   B) a + b = b + a
   C) a + 0 = a
   D) a + (-a) =0

4. $f(x) = 3x - 2; \ f^{-1}(x) =$

   A) $3x + 2$
   B) $x/6$
   C) $2x - 3$
   D) $(x + 2)/3$

5. What would be the total cost of a suit for $295.99 and a pair of shoes for $69.95 including 6.5% sales tax?

   A) $389.73
   B) $398.37
   C) $237.86
   D) $315.23

6. A student had 60 days to appeal the results of an exam. If the results were received on March 23, what was the last day that the student could appeal?

   A) May 21
   B) May 22
   C) May 23
   D) May 24

7. Which of the following is always composite if x is odd, y is even, and both x and y are greater than or equal to 2?

   A) $x + y$
   B) $3x + 2y$
   C) $5xy$
   D) $5x + 3y$

8. Which of the following is incorrect?

   A) $(x^2 y^3)^2 = x^4 y^6$
   B) $m^2 (2n)^3 = 8m^2 n^3$
   C) $(m^3 n^4)/(m^2 n^2) = mn^2$
   D) $(x + y^2)^2 = x^2 + y^4$

9. Express .0000456 in scientific notation.

    A) $4.56x10^{-4}$
    B) $45.6x10^{-6}$
    C) $4.56x10^{-6}$
    D) $4.56x10^{-5}$

10. Compute the area of the shaded region, given a radius of 5 meters. 0 is the center.

    A) 7.13 cm²
    B) 7.13 m²
    C) 78.5 m²
    D) 19.63 m²

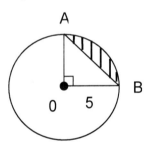

11. If the area of the base of a cone is tripled, the volume will be

    A) the same as the original
    B) 9 times the original
    C) 3 times the original
    D) 3 π times the original

12. Find the area of the figure pictured below.

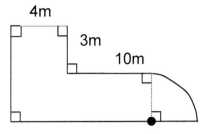

    A) 136.47 m²
    B) 148.48 m²
    C) 293.86 m²
    D) 178.47 m²1`

13. The mass of a Chips Ahoy cookie would be approximately equal to

    A) 1 kilogram
    B) 1 gram
    C) 15 grams
    D) 15 milligrams

14. Compute the median for the following data set:

    {12, 19, 13, 16, 17, 14}

    A) 14.5
    B) 15.17
    C) 15
    D) 16

15. Half the students in a class scored 80% on an exam, most of the rest scored 85% except for one student who scored 10%. Which would be the best measure of central tendency for the test scores?

A) mean
B) median
C) mode
D) either the median or the mode because they are equal

16. What conclusion can be drawn from the graph below?

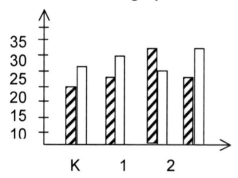

MLK Elementary
Student Enrollment   Girls   Boys

A) The number of students in first grade exceeds the number in second grade.
B) There are more boys than girls in the entire school.
C) There are more girls than boys in the first grade.
D) Third grade has the largest number of students.

17) State the domain of the function $f(x) = \dfrac{3x-6}{x^2-25}$

A) $x \neq 2$
B) $x \neq 5, -5$
C) $x \neq 2, -2$
D) $x \neq 5$

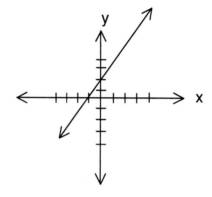

18. What is the equation of the above graph?

A) $2x + y = 2$
B) $2x - y = -2$
C) $2x - y = 2$
D) $2x + y = -2$

19. Solve for v_0 : $d = at(v_t - v_0)$

A) $v_0 = atd - v_t$
B) $v_0 = d - atv_t$
C) $v_0 = atv_t - d$
D) $v_0 = (atv_t - d)/at$

20. Which of the following is a factor of $6 + 48m^3$

A) (1 + 2m)
B) (1 - 8m)
C) (1 + m - 2m)
D) (1 - m + 2m)

21. Which graph represents the equation of $y = x^2 + 3x$?

A)   B)

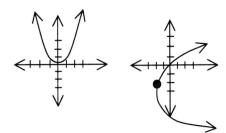

C)   D)

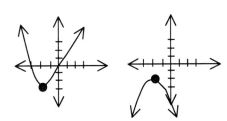

22. The volume of water flowing through a pipe varies directly with the square of the radius of the pipe. If the water flows at a rate of 80 liters per minute through a pipe with a radius of 4 cm, at what rate would water flow through a pipe with a radius of 3 cm?

A) 45 liters per minute
B) 6.67 liters per minute
C) 60 liters per minute
D) 4.5 liters per minute

23) Solve the system of equations for x, y and z.

$3x + 2y - z = 0$

$2x + 5y = 8z$

$x + 3y + 2z = 7$

A) $(-1, 2, 1)$
B) $(1, 2, -1)$
C) $(-3, 4, -1)$
D) $(0, 1, 2)$

24. Solve for x: $18 = 4 + |2x|$

A) $\{-11, 7\}$
B) $\{-7, 0, 7\}$
C) $\{-7, 7\}$
D) $\{-11, 11\}$

25. Which graph represents the solution set for $x^2 - 5x > -6$?

A) ![number line] -2 0 2

B) ![number line] -3 0

C) ![number line] -2 0 2

D) ![number line] -3 0 2 3

26. Find the zeroes of
$f(x) = x^3 + x^2 - 14x - 24$

A) 4, 3, 2
B) 3, -8
C) 7, -2, -1
D) 4, -3, -2

27. Evaluate $3^{1/2}(9^{1/3})$

   A) $27^{5/6}$
   B) $9^{7/12}$
   C) $3^{5/6}$
   D) $3^{6/7}$

28. Simplify: $\sqrt{27} + \sqrt{75}$

   A) $8\sqrt{3}$
   B) 34
   C) $34\sqrt{3}$
   D) $15\sqrt{3}$

29. Simplify: $\dfrac{10}{1+3i}$

   A) $-1.25(1-3i)$
   B) $1.25(1+3i)$
   C) $1+3i$
   D) $1-3i$

30. Find the sum of the first one hundred terms in the progression.
    (-6, -2, 2 . . .)

   A) 19,200
   B) 19,400
   C) -604
   D) 604

31. How many ways are there to choose a potato and two green vegetables from a choice of three potatoes and seven green vegetables?

   A) 126
   B) 63
   C) 21
   D) 252

32. What would be the seventh term of the expanded binomial $(2a+b)^8$?

   A) $2ab^7$
   B) $41a^4b^4$
   C) $112a^2b^6$
   D) $16ab^7$

33. Which term most accurately describes two coplanar lines without any common points?

   A) perpendicular
   B) parallel
   C) intersecting
   D) skew

34. Determine the number of subsets of set K.
    K = {4, 5, 6, 7}

   A) 15
   B) 16
   C) 17
   D) 18

35. What is the degree measure of each interior angle of a regular 10 sided polygon?

   A) 18°
   B) 36°
   C) 144°
   D) 54°

36. If a ship sails due south 6 miles, then due west 8 miles, how far was it from the starting point?

   A) 100 miles
   B) 10 miles
   C) 14 miles
   D) 48 miles

37. What is the measure of minor arc AD, given measure of arc PS is 40° and $m < K = 10°$?

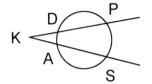

   A) 50°
   B) 20°
   C) 30°
   D) 25°

38. Choose the diagram which illustrates the construction of a perpendicular to the line at a given point on the line.

   A)

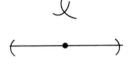

   B)

   C)

   D)

39. When you begin by assuming the conclusion of a theorem is false, then show that through a sequence of logically correct steps you contradict an accepted fact, this is known as

   A) inductive reasoning
   B) direct proof
   C) indirect proof
   D) exhaustive proof

40. Which theorem can be used to prove $\triangle BAK \cong \triangle MKA$?

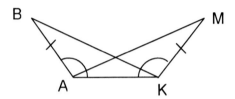

   A) SSS
   B) ASA
   C) SAS
   D) AAS

41. Given that QO⊥NP and QO=NP, quadrilateral NOPQ can most accurately be described as a

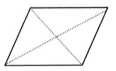

   A) parallelogram
   B) rectangle
   C) square
   D) rhombus

42. Choose the correct statement concerning the median and altitude in a triangle.

   A) The median and altitude of a triangle may be the same segment.
   B) The median and altitude of a triangle are always different segments.
   C) The median and altitude of a right triangle are always the same segment.
   D) The median and altitude of an isosceles triangle are always the same segment.

43. Which mathematician is best known for his work in developing non-Euclidean geometry?

   A) Descartes
   B) Riemann
   C) Pascal
   D) Pythagoras

44. Find the surface area of a box which is 3 feet wide, 5 feet tall, and 4 feet deep.

   A) 47 sq. ft.
   B) 60 sq. ft.
   C) 94 sq. ft
   D) 188 sq. ft.

45. Given a 30 meter x 60 meter garden with a circular fountain with a 5 meter radius, calculate the area of the portion of the garden not occupied by the fountain.

   A) 1721 m²
   B) 1879 m²
   C) 2585 m²
   D) 1015 m²

46. Determine the area of the shaded region of the trapezoid in terms of x and y.

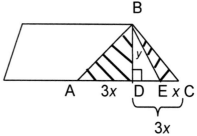

   A) $4xy$
   B) $2xy$
   C) $3x^2 y$
   D) There is not enough information given.

47. Compute the standard deviation for the following set of temperatures.
(37, 38, 35, 37, 38, 40, 36, 39)

   A) 37.5
   B) 1.5
   C) 0.5
   D) 2.5

48. **Find the value of the determinant of the matrix.**

$$\begin{vmatrix} 2 & 1 & -1 \\ 4 & -1 & 4 \\ 0 & -3 & 2 \end{vmatrix}$$

A) 0
B) 23
C) 24
D) 40

49. **Which expression is equivalent to $1 - \sin^2 x$?**

A) $1 - \cos^2 x$
B) $1 + \cos^2 x$
C) $1 / \sec x$
D) $1 / \sec^2 x$

50. **Determine the measures of angles A and B.**

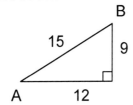

A) A = 30°, B = 60°
B) A = 60°, B = 30°
C) A = 53°, B = 37°
D) A = 37°, B = 53°

51. **Given $f(x) = 3x - 2$ and $g(x) = x^2$, determine $g(f(x))$.**

A) $3x^2 - 2$
B) $9x^2 + 4$
C) $9x^2 - 12x + 4$
D) $3x^3 - 2$

52. **Determine the rectangular coordinates of the point with polar coordinates (5, 60°).**

A) (0.5, 0.87)
B) (-0.5, 0.87)
C) (2.5, 4.33)
D) (25, 150°)

53. **Given a vector with horizontal component 5 and vertical component 6, determine the length of the vector.**

A) 61
B) $\sqrt{61}$
C) 30
D) $\sqrt{30}$

54. **Compute the distance from (-2,7) to the line $x = 5$.**

A) -9
B) -7
C) 5
D) 7

55. **Given $K(-4, y)$ and $M(2, -3)$ with midpoint $L(x, 1)$, determine the values of x and y.**

A) $x = -1, y = 5$
B) $x = 3, y = 2$
C) $x = 5, y = -1$
D) $x = -1, y = -1$

56. **Find the length of the major axis of** $x^2 + 9y^2 = 36$.

A) 4
B) 6
C) 12
D) 8

57. **Which equation represents a circle with a diameter whose endpoints are** $(0,7)$ **and** $(0,3)$**?**

A) $x^2 + y^2 + 21 = 0$
B) $x^2 + y^2 - 10y + 21 = 0$
C) $x^2 + y^2 - 10y + 9 = 0$
D) $x^2 - y^2 - 10y + 9 = 0$

58. **Which equation corresponds to the logarithmic statement:** $\log_x k = m$**?**

A) $x^m = k$
B) $k^m = x$
C) $x^k = m$
D) $m^x = k$

59. **Find the first derivative of the function:**
$f(x) = x^3 - 6x^2 + 5x + 4$

A) $3x^3 - 12x^2 + 5x = f'(x)$
B) $3x^2 - 12x - 5 = f'(x)$
C) $3x^2 - 12x + 9 = f'(x)$
D) $3x^2 - 12x + 5 = f'(x)$

60. **Differentiate:** $y = e^{3x+2}$

A) $3e^{3x+2} = y'$
B) $3e^{3x} = y'$
C) $6e^3 = y'$
D) $(3x+2)e^{3x+1} = y'$

61. **Find the slope of the line tangent to** $y = 3x(\cos x)$ **at** $(\pi/2,\ \pi/2)$**.**

A) $-3\pi/2$
B) $3\pi/2$
C) $\pi/2$
D) $-\pi/2$

62. **Find the equation of the line tangent to** $y = 3x^2 - 5x$ **at** $(1,-2)$**.**

A) $y = x - 3$
B) $y = 1$
C) $y = x + 2$
D) $y = x$

63. **How does the function** $y = x^3 + x^2 + 4$ **behave from** $x = 1$ **to** $x = 3$**?**

A) increasing, then decreasing
B) increasing
C) decreasing
D) neither increasing nor decreasing

64. Find the absolute maximum obtained by the function $y = 2x^2 + 3x$ on the interval $x = 0$ to $x = 3$.

    A) $-3/4$
    B) $-4/3$
    C) 0
    D) 27

65. Find the antiderivative for $4x^3 - 2x + 6 = y$.

    A) $x^4 - x^2 + 6x + C$
    B) $x^4 - 2/3x^3 + 6x + C$
    C) $12x^2 - 2 + C$
    D) $\frac{4}{3}x^4 - x^2 + 6x + C$

66. Find the antiderivative for the function $y = e^{3x}$.

    A) $3x(e^{3x}) + C$
    B) $3(e^{3x}) + C$
    C) $1/3(e^x) + C$
    D) $1/3(e^{3x}) + C$

67. The acceleration of a particle is dv/dt = 6 m/s². Find the velocity at t=10 given an initial velocity of 15 m/s.

    A) 60 m/s
    B) 150 m/s
    C) 75 m/s
    D) 90 m/s

68. If the velocity of a body is given by v = 16 - t², find the distance traveled from t = 0 until the body comes to a complete stop.

    A) 16
    B) 43
    C) 48
    D) 64

69. Evaluate: $\int (x^3 + 4x - 5)dx$

    A) $3x^2 + 4 + C$
    B) $\frac{1}{4}x^4 + 2x^2 - 5x + C$
    C) $x^{4/3} + 4x - 5x + C$
    D) $x^3 + 4x^2 - 5x + C$

70. Evaluate $\int_0^2 (x^2 + x - 1)dx$

    A) 11/3
    B) 8/3
    C) -8/3
    D) -11/3

71. Find the area under the function $y = x^2 + 4$ from $x = 3$ to $x = 6$.

    A) 75
    B) 21
    C) 96
    D) 57

72. -3 + 7 = -4   6(-10) = - 60
     -5(-15) = 75   -3+-8 = 11
     8-12 = -4   7- -8 = 15

Which best describes the type of error observed above?

A) The student is incorrectly multiplying integers.
B) The student has incorrectly applied rules for adding integers to subtracting integers.
C) The student has incorrectly applied rules for multiplying integers to adding integers.
D) The student is incorrectly subtracting integers.

73. **About two weeks after introducing formal proofs, several students in your geometry class are having a difficult time remembering the names of the postulates. They cannot complete the reason column of the proof and as a result are not even attempting the proofs. What would be the best approach to help students understand the nature of geometric proofs?**

A) Give them more time; proofs require time and experience.
B) Allow students to write an explanation of the theorem in the reason column instead of the name.
C) Have the student copy each theorem in a notebook.
D) Allow the students to have open book tests.

74. **What would be the shortest method of solution for the system of equations below?**

$$3x + 2y = 38$$
$$4x + 8 = y$$

A) linear combination
B) additive inverse
C) substitution
D) graphing

75. **Identify the correct sequence of subskills required for solving and graphing inequalities involving absolute value in one variable, such as $|x+1| \leq 6$.**

A) understanding absolute value, graphing inequalities, solving systems of equations
B) graphing inequalities on a Cartesian plane, solving systems of equations, simplifying expressions with absolute value
C) plotting points, graphing equations, graphing inequalities
D) solving equations with absolute value, solving inequalities, graphing conjunctions and disjunctions

76. **What would be the least appropriate use for handheld calculators in the classroom?**

A) practice for standardized tests
B) integrating algebra and geometry with applications
C) justifying statements in geometric proofs
D) applying the law of sines to find dimensions

77. **According to Piaget, what stage in a student's development would be appropriate for introducing abstract concepts in geometry?**

A) concrete operational
B) formal operational
C) sensori-motor
D) pre-operational

78. **A group of students working with trigonometric identities have concluded that $\cos 2x = 2\cos x$. How could you best lead them to discover their error?**

A) Have the students plug in values on their calculators.
B) Direct the student to the appropriate chapter in the text.
C) Derive the correct identity on the board.
D) Provide each student with a table of trig identities.

79. **Which of the following is the best example of the value of personal computers in advanced high school mathematics?**

A) Students can independently drill and practice test questions.
B) Students can keep an organized list of theorems and postulates on a word processing program.
C) Students can graph and calculate complex functions to explore their nature and make conjectures.
D) Students are better prepared for business because of mathematics computer programs in high school.

80. **Given the series of examples below, what is 5¢4?**

4¢3=13   7¢2=47
3¢1=8   1¢5=-4

A) 20
B) 29
C) 1
D) 21

Answer Key

1)	B	17)	B	33)	B	49)	D	65)	A
2)	B	18)	B	34)	B	50)	D	66)	D
3)	D	19)	D	35)	C	51)	C	67)	C
4)	D	20)	A	36)	B	52)	C	68)	B
5)	A	21)	C	37)	B	53)	B	69)	B
6)	B	22)	A	38)	D	54)	D	70)	B
7)	C	23)	A	39)	C	55)	A	71)	A
8)	D	24)	C	40)	C	56)	C	72)	C
9)	D	25)	D	41)	C	57)	B	73)	B
10)	B	26)	D	42)	A	58)	A	74)	C
11)	C	27)	B	43)	B	59)	D	75)	D
12)	B	28)	A	44)	C	60)	A	76)	C
13)	C	29)	D	45)	A	61)	A	77)	B
14)	C	30)	A	46)	B	62)	C	78)	C
15)	B	31)	A	47)	B	63)	B	79)	C
16)	B	32)	C	48)	C	64)	D	80)	D

Rationales for Sample Questions

1. Let N = .636363.... Then multiplying both sides of the equation by 100 or 10^2 (because there are 2 repeated numbers), we get 100N = 63.636363... Then subtracting the two equations gives 99N = 63 or N = $\dfrac{63}{99}$ = $\dfrac{7}{11}$. **Answer is B**

2. I is not closed because $\dfrac{4}{.5} = 8$ and 8 is not in the set.

   III is not closed because $\dfrac{1}{0}$ is undefined.

   II is closed because $\dfrac{-1}{1} = -1, \dfrac{1}{-1} = -1, \dfrac{1}{1} = 1, \dfrac{-1}{-1} = 1$ and all the answers are in the set. **Answer is B**

3. **Answer is D** because a + (-a) = 0 is a statement of the Additive Inverse Property of Algebra.

4. To find the inverse, $f^{-1}(x)$, of the given function, reverse the variables in the given equation, y = 3x – 2, to get x = 3y – 2. Then solve for y as follows: x+2 = 3y, and y = $\dfrac{x+2}{3}$. **Answer is D.**

5. Before the tax, the total comes to $365.94. Then .065(365.94) = 23.79. With the tax added on, the total bill is 365.94 + 23.79 = $389.73. (Quicker way: 1.065(365.94) = 389.73.) **Answer is A**

6. Recall: 30 days in April and 31 in March. 8 days in March + 30 days in April + 22 days in May brings him to a total of 60 days on May 22. **Answer is B.**

7. A composite number is a number which is not prime. The prime number sequence begins 2,3,5,7,11,13,17,.... To determine which of the expressions is <u>always</u> composite, experiment with different values of x and y, such as x=3 and y=2, or x=5 and y=2. It turns out that 5xy will always be an even number, and therefore, composite, if y=2.
Answer is C.

8. Using FOIL to do the expansion, we get $(x + y^2)^2 = (x + y^2)(x + y^2) = x^2 + 2xy^2 + y^4$.
Answer is D.

9. In scientific notation, the decimal point belongs to the right of the 4, the first significant digit. To get from 4.56×10^{-5} back to 0.0000456, we would move the decimal point 5 places to the left.
Answer is D.

10. Area of triangle AOB is .5(5)(5) = 12.5 square meters. Since $\frac{90}{360} = .25$, the area of sector AOB (pie-shaped piece) is approximately $.25(\pi)5^2 = 19.63$. Subtracting the triangle area from the sector area to get the area of segment AB, we get approximately 19.63-12.5 = 7.13 square meters. **Answer is B.**

11. The formula for the volume of a cone is V = $\frac{1}{3}Bh$, where B is the area of the circular base and h is the height. If the area of the base is tripled, the volume becomes

V = $\frac{1}{3}(3B)h = Bh$, or three times the original area. **Answer is C.**

12. Divide the figure into 2 rectangles and one quarter circle. The tall rectangle on the left will have dimensions 10 by 4 and area 40. The rectangle in the center will have dimensions 7 by 10 and area 70. The quarter circle will have area $.25(\pi)7^2 = 38.48$.
The total area is therefore approximately 148.48. **Answer is B.**

13. Since an ordinary cookie would not weigh as much as 1 kilogram, or as little as 1 gram or 15 milligrams, the only reasonable answer is 15 grams. **Answer is C.**

14. Arrange the data in ascending order: 12,13,14,16,17,19. The median is the middle value in a list with an odd number of entries. When there is an even number of entries, the median is the mean of the two center entries. Here the average of 14 and 16 is 15. **Answer is C.**

15. In this set of data, the median (see #14) would be the most representative measure of central tendency since the median is independent of extreme values. Because of the 10% outlier, the mean (average) would be disproportionately skewed. In this data set, it is true that the median and the mode (number which occurs most often) are the same, but the median remains the best choice because of its special properties. **Answer is B.**

16. In kindergarten, first grade, and third grade, there are more boys than girls. The number of extra girls in grade two is more than made up for by the extra boys in all the other grades put together. **Answer is B.**

17. The values of 5 and –5 must be omitted from the domain of all real numbers because if x took on either of those values, the denominator of the fraction would have a value of 0, and therefore the fraction would be undefined. **Answer is B.**

18. By observation, we see that the graph has a y-intercept of 2 and a slope of 2/1 = 2. Therefore its equation is y = mx + b = 2x + 2. Rearranging the terms gives
2x – y = -2. **Answer is B**.

19. Using the Distributive Property and other properties of equality to isolate v_0 gives
d = atv_t – atv_0, atv_0 = atv_t – d, $v_0 = \dfrac{atv_t - d}{at}$. **Answer is D.**

20. Removing the common factor of 6 and then factoring the sum of two cubes gives
$6 + 48m^3 = 6(1 + 8m^3) = 6(1 + 2m)(1^2 - 2m + (2m)^2)$. **Answer is A**.

21. B is not the graph of a function. D is the graph of a parabola where the coefficient of x^2 is negative. A appears to be the graph of $y = x^2$. To find the x-intercepts of
$y = x^2 + 3x$, set y = 0 and solve for x: $0 = x^2 + 3x = x(x + 3)$ to get x = 0 or x = -3. Therefore, the graph of the function intersects the x-axis at x=0 and x=-3.
Answer is C.

22. Set up the direct variation: $\dfrac{V}{r^2} = \dfrac{V}{r^2}$. Substituting gives $\dfrac{80}{16} = \dfrac{V}{9}$. Solving for V gives 45 liters per minute. **Answer is A.**

23. Multiplying equation 1 by 2, and equation 2 by –3, and then adding together the two resulting equations gives -11y + 22z = 0. Solving for y gives y = 2z. In the meantime, multiplying equation 3 by –2 and adding it to equation 2 gives
–y – 12z = -14. Then substituting 2z for y, yields the result z = 1. Subsequently, one can easily find that y = 2, and x = -1. **Answer is A.**

24. Using the definition of absolute value, two equations are possible: 18 = 4 + 2x or 18 = 4 – 2x. Solving for x gives x = 7 or x = -7. **Answer is C**.

25. Rewriting the inequality gives $x^2 - 5x + 6 > 0$. Factoring gives (x – 2)(x – 3) > 0.
The two cut-off points on the numberline are now at x = 2 and x = 3. Choosing a random number in each of the three parts of the numberline, we test them to see if they produce a true statement. If x = 0 or x = 4, (x-2)(x-3)>0 is true. If x = 2.5, (x-2)(x-3)>0 is false. Therefore the solution set is all numbers smaller than 2 or greater than 3.
Answer is D.

26. Possible rational roots of the equation $0 = x^3 + x^2 - 14x - 24$ are all the positive and negative factors of 24. By substituting into the equation, we find that -2 is a root, and therefore that $x+2$ is a factor. By performing the long division $(x^3 + x^2 - 14x - 24)/(x+2)$, we can find that another factor of the original equation is $x^2 - x - 12$ or $(x-4)(x+3)$. Therefore the zeros of the original function are -2, -3, and 4. **Answer is D.**

27. Getting the bases the same gives us $3^{\frac{1}{2}}3^{\frac{2}{3}}$. Adding exponents gives $3^{\frac{7}{6}}$. Then some additional manipulation of exponents produces $3^{\frac{7}{6}} = 3^{\frac{14}{12}} = (3^2)^{\frac{7}{12}} = 9^{\frac{7}{12}}$. **Answer is B.**

28. Simplifying radicals gives $\sqrt{27} + \sqrt{75} = 3\sqrt{3} + 5\sqrt{3} = 8\sqrt{3}$. **Answer is A.**

29. Multiplying numerator and denominator by the conjugate gives $\frac{10}{1+3i} \times \frac{1-3i}{1-3i} = \frac{10(1-3i)}{1-9i^2} = \frac{10(1-3i)}{1-9(-1)} = \frac{10(1-3i)}{10} = 1 - 3i$. **Answer is D.**

30. To find the 100th term: $t_{100} = -6 + 99(4) = 390$. To find the sum of the first 100 terms: $S = \frac{100}{2}(-6 + 390) = 19200$. **Answer is A.**

31. There are 3 slots to fill. There are 3 choices for the first, 7 for the second, and 6 for the third. Therefore, the total number of choices is $3(7)(6) = 126$. **Answer is A.**

32. The set-up for finding the seventh term is $\frac{8(7)(6)(5)(4)(3)}{6(5)(4)(3)(2)(1)}(2a)^{8-6}b^6$ which gives

$28(4a^2b^6)$ or $112a^2b^6$. **Answer is C.**

33. By definition, parallel lines are coplanar lines without any common points. **Answer is B.**

34. A set of n objects has 2^n subsets. Therefore, here we have $2^4 = 16$ subsets. These subsets include four which have only 1 element each, six which have 2 elements each, four which have 3 elements each, plus the original set, and the empty set. **Answer is B.**

35. Formula for finding the measure of each interior angle of a regular polygon with n sides is $\frac{(n-2)180}{n}$. For n=10, we get $\frac{8(180)}{10} = 144$. **Answer is C.**

36. Draw a right triangle with legs of 6 and 8. Find the hypotenuse using the Pythagorean Theorem. $6^2 + 8^2 = c^2$. Therefore, c = 10 miles. **Answer is B.**

37. The formula relating the measure of angle K and the two arcs it intercepts is $m\angle K = \frac{1}{2}(mPS - mAD)$. Substituting the known values, we get $10 = \frac{1}{2}(40 - mAD)$. Solving for mAD gives an answer of 20 degrees. **Answer is B.**

38. Given a point on a line, place the compass point there and draw two arcs intersecting the line in two points, one on either side of the given point. Then using any radius larger than half the new segment produced, and with the pointer at each end of the new segment, draw arcs which intersect above the line. Connect this new point with the given point. **Answer is D.**

39. By definition this describes the procedure of an indirect proof. **Answer is C.**

40. Since side AK is common to both triangles, the triangles can be proved congruent by using the Side-Angle-Side Postulate. **Answer is C.**

41. In an ordinary parallelogram, the diagonals are not perpendicular or equal in length. In a rectangle, the diagonals are not necessarily perpendicular. In a rhombus, the diagonals are not equal in length. In a square, the diagonals are both perpendicular and congruent. **Answer is C.**

42. The most one can say with certainty is that the median (segment drawn to the midpoint of the opposite side) and the altitude (segment drawn perpendicular to the opposite side) of a triangle <u>may</u> coincide, but they more often do not. In an isosceles triangle, the median and the altitude to the <u>base</u> are the same segment. **Answer is A.**

43. In the mid-nineteenth century, Reimann and other mathematicians developed elliptic geometry. **Answer is B.**

44. Let's assume the base of the rectangular solid (box) is 3 by 4, and the height is 5. Then the surface area of the top and bottom together is 2(12) = 24. The sum of the areas of the front and back are 2(15) = 30, while the sum of the areas of the sides are 2(20)=40. The total surface area is therefore 94 square feet. **Answer is C.**

45. Find the area of the garden and then subtract the area of the fountain: $30(60) - \pi(5)^2$ or approximately 1721 square meters. **Answer is A.**

46. To find the area of the shaded region, find the area of triangle ABC and then subtract the area of triangle DBE. The area of triangle ABC is .5(6x)(y) = 3xy. The area of triangle DBE is .5(2x)(y) = xy. The difference is 2xy. **Answer is B.**

47. Find the mean: 300/8 = 37.5. Then, using the formula for standard deviation, we get

$$\sqrt{\frac{2(37.5-37)^2 + 2(37.5-38)^2 + (37.5-35)^2 + (37.5-40)^2 + (37.5-36)^2 + (37.5-39)^2}{8}}$$

which has a value of 1.5. **Answer is B.**

48. To find the determinant of a matrix without the use of a graphing calculator, repeat the first two columns as shown,

2	1	-1	2	1
4	-1	4	4	-1
0	-3	2	0	-3

Starting with the top left-most entry, 2, multiply the three numbers in the diagonal going down to the right: 2(-1)(2)=-4. Do the same starting with 1: 1(4)(0)=0. And starting with –1: -1(4)(-3) = 12. Adding these three numbers, we get 8. Repeat the same process starting with the top right-most entry, 1. That is, multiply the three numbers in the diagonal going down to the left: 1(4)(2) = 8. Do the same starting with 2: 2(4)(-3) = -24 and starting with –1: -1(-1)(0) = 0. Add these together to get
-16. To find the determinant, subtract the second result from the first: 8-(-16)=24.
Answer is C.

49. Using the Pythagorean Identity, we know $\sin^2 x + \cos^2 x = 1$. Thus $1 - \sin^2 x = \cos^2 x$, which by definition is equal to $1/\sec^2 x$. **Answer is D.**

50. Tan A = 9/12=.75 and $\tan^{-1}.75$ = 37 degrees. Since angle B is complementary to angle A, the measure of angle B is therefore 53 degrees. **Answer is D.**

51. The composite function $g(f(x)) = (3x-2)^2 = 9x^2 - 12x + 4$. **Answer is C.**

52. Given the polar point $(r, \theta) = (5, 60)$, we can find the rectangular coordinates this way: $(x,y) = (r\cos\theta, r\sin\theta) = (5\cos 60, 5\sin 60) = (2.5, 4.33)$. **Answer is C.**

53. Using the Pythagorean Theorem, we get $v = \sqrt{36+25} = \sqrt{61}$. **Answer is B.**

54. The line x = 5 is a vertical line passing through (5,0) on the Cartesian plane. By observation the distance along the horizontal line from the point (-2,7) to the line x=5 is 7 units. **Answer is D.**

55. The formula for finding the midpoint (a,b) of a segment passing through the points $(x_1, y_1) and (x_2, y_2) is (a,b) = (\frac{x_1 + x_2}{2}, \frac{y_1 + y_2}{2})$. Setting up the corresponding equations from this information gives us $x = \frac{-4 + 2}{2}, and 1 = \frac{y - 3}{2}$. Solving for x and y gives
x = -1 and y = 5. **Answer is A**.

56. Dividing by 36, we get $\frac{x^2}{36} + \frac{y^2}{4} = 1$, which tells us that the ellipse intersects the x-axis at 6 and –6, and therefore the length of the major axis is 12. (The ellipse intersects the y-axis at 2 and –2). **Answer is C**.

57. With a diameter going from (0,7) to (0,3), the diameter of the circle must be 4, the radius must be 2, and the center of the circle must be at (0,5). Using the standard form for the equation of a circle, we get $(x-0)^2 + (y-5)^2 = 2^2$. Expanding, we get
$x^2 + y^2 - 10y + 21 = 0$. **Answer is B**.

58. By definition of log form and exponential form, $\log_x k = m$ corresponds to x^m = k. **Answer is A**.

59. Use the Power Rule for polynomial differentiation: if $y = ax^n$, then $y' = nax^{n-1}$. **Answer is D**.

60. Use the Exponential Rule for derivatives of functions of e: if $y = ae^{f(x)}$, then $y' = f'(x)ae^{f(x)}$. **Answer is A**.

61. To find the slope of the tangent line, find the derivative, and then evaluate it at
$x = \frac{\pi}{2}$. y' = 3x(-sinx)+3cosx. At the given value of x,
$y' = 3(\frac{\pi}{2})(-\sin\frac{\pi}{2}) + 3\cos\frac{\pi}{2} = \frac{-3\pi}{2}$. **Answer is A**.

62. To find the slope of the tangent line, find the derivative, and then evaluate it at x=1.
y'=6x-5=6(1)-5=1. Then using point-slope form of the equation of a line, we get y+2=1(x-1) or y = x-3. **Answer is C**.

63. To find critical points, take the derivative, set it equal to 0, and solve for x. f'(x) = $3x^2$ + 2x = x(3x+2)=0. CP at x=0 and x=-2/3. Neither of these CP is on the interval from x=1 to x=3. Testing the endpoints: at x=1, y=6 and at x=3, y=38. Since the derivative is positive for all values of x from x=1 to x=3, the curve is increasing on the entire interval. **Answer is B**.

64. Find CP at x=-.75 as done in #63. Since the CP is not in the interval from x=0 to x=3, just find the values of the functions at the endpoints. When x=0, y=0, and when x=3, y = 27. Therefore 27 is the absolute maximum on the given interval.
Answer is D.

65. Use the rule for polynomial integration: given ax^n, the antiderivative is $\frac{ax^{n+1}}{n+1}$. **Answer is A.**

66. Use the rule for integration of functions of e: $e^x dx = e^x + C$. **Answer is D.**

67. Recall that the derivative of the velocity function is the acceleration function. In reverse, the integral of the acceleration function is the velocity function. Therefore, if a=6, then v=6t+C. Given that at t=0, v=15, we get v = 6t+15. At t=10, v=60+15=75m/s. **Answer is C.**

68. Recall that the derivative of the distance function is the velocity function. In reverse, the integral of the velocity function is the distance function. To find the time needed for the body to come to a stop when v=0, solve for t: $v = 16 - t^2 = 0$.
Result: t = 4 seconds. The distance function is s = 16t - $\frac{t^3}{3}$. At t=4, s= 64 – 64/3 or approximately 43 units. **Answer is B.**

69. Integrate as described in #65. **Answer is B.**

70. Use the fundamental theorem of calculus to find the definite integral: given a continuous function f on an interval [a,b], then $\int_a^b f(x)dx = F(b) - F(a)$, where F is an antiderivative of f.

$\int_0^2 (x^2 + x - 1)dx = (\frac{x^3}{3} + \frac{x^2}{2} - x)$ Evaluate the expression at x=2, at x=0, and then subtract to get 8/3 + 4/2 – 2-0 = 8/3. **Answer is B.**

71. To find the area set up the definite integral: $\int_3^6 (x^2 + 4)dx = (\frac{x^3}{3} + 4x)$.
Evaluate the expression at x=6, at x=3, and then subtract to get (72+24)-(9+12)=75.
Answer is A.

72. The errors are in the following: -3+7=-4 and –3 + -8 = 11, where the student seems to be using the rules for signs when multiplying, instead of the rules for signs when adding. **Answer is C.**

73. **Answer is B.**

74. Since the second equation is already solved for y, it would be easiest to use the substitution method. **Answer is C.**

75. The steps listed in answer D would look like this for the given example: If $|x + 1| \le 6$, then $-6 \le x + 1 \le 6$, which means $-7 \le x \le 5$. Then the inequality would be graphed on a numberline and would show that the solution set is all real numbers between −7 and 5, including −7 and 5. **Answer is D.**

76. There is no need for calculators when justifying statements in a geometric proof.
Answer is C.

77. By observation, the **Answer is B.**

78. **Answer is C.**

79. **Answer is C.**

80. By observation of the examples given, $a \not\subset b = a^2 - b$. Therefore, $5 \not\subset 4 = 25 - 4 = 21$. **Answer is D.**

Good luck to all of you. May you have many happy years teaching. The end.

XAMonline, INC. 21 Orient Ave. Melrose, MA 02176

Toll Free number 800-509-4128

TO ORDER Fax 781-662-9268 OR www.XAMonline.com

CERTIFICATION EXAMINATION FOR OKLAHOMA EDUCATORS - CEOE - 2008

PO#   Store/School:

Address 1:

Address 2 (Ship to other):

City, State Zip

Credit card number_____-_____-_____-_____   expiration_____

EMAIL _____

PHONE   FAX

ISBN	TITLE	Qty	Retail	Total
978-1-58197-775-2	CEOE OSAT Art Sample Test Field 02		$15.00	
978-1-58197-776-9	CEOE OSAT Chemistry Field 04		$59.95	
978-1-58197-777-6	CEOE OSAT English Field 07		$59.95	
978-1-58197-778-3	CEOE OSAT Earth Science Field 08		$59.95	
978-1-58197-780-6	CEOE OSAT Biological Sciences Field 10		$59.95	
978-1-58197-646-5	CEOE OSAT Advanced Mathematics Field 11		$59.95	
978-1-58197-782-0	CEOE OSAT Physical Education-Health-Safety		$59.95	
978-1-58197-663-2	CEOE OSAT Physics Field 14		$59.95	
978-1-58197-784-4	CEOE OSAT Reading Specialist Field 15		$59.95	
978-1-58197-785-1	CEOE OSAT Spanish Field 19		$59.95	
978-1-58197-786-8	CEOE OSAT French Sample Test Field 20		$15.00	
978-1-58197-787-5	CEOE OSAT Middle Level English Field 24		$59.95	
978-1-58197-647-2	CEOE OSAT Middle Level-Intermediate Mathematics Field 25		$59.95	
978-1-58197-789-9	CEOE OSAT Middle Level Science Field 26		$59.95	
978-1-58197-790-5	CEOE OSAT Middle Level Social Studies Field 27		$59.95	
978-1-58197-791-2	CEOE OSAT Mild Moderate Disabilities Field 29		$73.50	
978-1-58197-792-9	CEOE OSAT Library-Media Specialist Field 38		$59.95	
978-1-58197-793-6	CEOE OSAT Principal Common Core Field 44		$59.95	
978-1-58197-794-3	CEOE OSAT Elementary Education Fields 50-51		$28.95	
978-1-58197-797-4	CEOE OSAT U.S. & World History Fields 17-18		$59.95	
978-1-58197-798-1	CEOE OGET Oklahoma General Education Test		$28.95	
978-1-58197-796-7	CEOE OPTE Oklahoma Professional Teaching Examination Fields 75-76		$59.95	
			SUBTOTAL	
			Ship	$8.25
			TOTAL	

CPSIA information can be obtained at www.ICGtesting.com
Printed in the USA
BVOW080324140212

282881BV00004B/33/P

9 781581 976465